A TASTE OF MOLECULES

THE WOMEN WRITING SCIENCE PROJECT

In collaboration with the National Science Foundation, the Feminist Press announces Women Writing Science, a series of books that bring to life the story of women in science, engineering, technology, and mathematics. From biographies and novels to research-based nonfiction, Women Writing Science will celebrate the achievements of women scientists, explore their conflicts, address gender bias barriers, encourage women and men of all ages to become more scientifically literate, and embolden young women to choose careers in science.

Other books in the series

WOMEN IN SCIENCE: THEN AND NOW
Vivian Gornick

BASE TEN
Maryann Lesert

THE MADAME CURIE COMPLEX: THE HIDDEN HISTORY OF WOMEN IN SCIENCE
Julie Des Jardins

A TASTE OF MOLECULES

In Search of the Secrets of Flavor

DIANE FRESQUEZ

Published in 2013 by the Feminist Press
at the City University of New York
The Graduate Center
365 Fifth Avenue, Suite 5406
New York, NY 10016

feministpress.org

Publication of this volume is made possible, in part, by funds from the National Science Foundation.

First printing October 2013

Cover and text design by Drew Stevens

Library of Congress Cataloging-in-Publication Data
Fresquez, Diane
A taste of molecules : in search of the secrets of flavor / by Diane Fresquez.
p. cm.
ISBN 978-1-55861-839-8 (pbk.)
1. Taste buds. 2. Flavor. 3. Gender identity in science.
4. Flavoring essences. I. Title.
QP456.F74 2013
612.8'7—dc23

2011052067

For my husband, Richard,
and our daughters, Louisa and Clara

CONTENTS

I. TASTE UNCORKED

L'AMUSE-BOUCHE

If you want to make an apple pie from scratch,
first you must invent the universe.
CARL SAGAN

Although it's often said that every journey begins with a single step, in this case it began with a single salsa. The sauce, not the dance. That is, half a dozen ripe tomatoes, a sprinkle of sugar (to improve the flavor of the tomatoes; omit if homegrown), two scallions, three cloves of garlic, one jalapeño pepper, and a handful of fresh coriander—everything chopped and stirred together in a bowl with a good splash of white vinegar and salt to taste.

It was a Sunday afternoon in October and my husband and I had been invited to our friends' house for a drink. We were late, I was distracted and disoriented, but at the last minute I decided to make salsa to bring along. I had just returned from a difficult trip home to the southwestern United States, where I was born and raised. My husband and I had lived in London for many years, but were now living in Brussels, a delightful place to which we had moved for work, along with our two daughters.

Belgium is about the size of Maryland, and although

small compared to its next-door neighbors, France and Germany, it produces more than 172,000 tons of chocolate a year, and is densely populated with excellent restaurants. It's also home to many an opinionated food lover, including a certain Renaissance man named Hughes Belin, a French journalist specializing in energy and climate issues, part-time restaurant critic, gourmand, and amateur singer (jazz and rock). He was our friends' good chum, and my husband and I met him for the first time at their home that afternoon. Having a drink with him really cheered me up. Hughes was even later than we were, and dressed casually with some of his top shirt buttons undone, exposing more chest hair than one usually sees on men nowadays. It was his signature style, apparently, and I don't think he would mind me mentioning it. On the Slow Food Bruxelles website there's a large photo of him wearing nothing but an apron and a cheeky grin.

Hughes sat on the edge of our friend Jacky's sofa as if ready to fly off at any minute to the next exciting rendezvous. He had that mercurial air of a bad-boy television chef, and talked animatedly about the book he was writing, a bilingual (French and English) dining-out guide to the city's restaurant-dense European Quarter, home of the European Commission and the European Parliament. He also talked about the challenge and the fun of cooking up something delicious with friends even though, and especially if, the cupboard is nearly bare. I was relieved we were at Jacky's and not my house, when I heard that he heads straight to the kitchen whenever he arrives at the home of friends, family or acquaintances.

"I open the fridge first," he said, and indeed I had seen him rummaging inside Jacky's when he arrived. Hughes was a man possessed, a foodie force of nature. Discovering nothing more than a fresh baguette and a piece of artisanal cheese in someone's kitchen could fill him with ineffable happiness, but finding a cupboard stuffed with packages of junk food could leave him in despair.

As Hughes talked, he ignored the small bowl of salsa sitting on the coffee table in front of him, and I became increasingly embarrassed that I had made it and brought it over in the first place. I had forgotten to bring a serving bowl, and it didn't look particularly appetizing sitting there in the small, glass mixing bowl I had used to transport it. Jacky prompted him to eat the salsa and he began absentmindedly, with no interest at first, until suddenly I saw his expression change to one of surprise and delight. Apparently it tasted far better than it looked. "My father puts the jalapeño in boiling water for a few mintues before chopping to bring out the flavor," I told Hughes. Thus, this small, piquant offering, my father's recipe, turned out to be a direct and lasting way into this restaurant critic's heart.

I was born and raised in New Mexico and had recently returned from a trip home to be at my elderly mother's bedside before she died. Now I was using salsa and other dishes, such as her recipe for arroz con pollo, which I was obsessed with cooking at the moment, as a way of mourning. I was lucky enough to have brought back with me a good supply of the ingredient that gave this simple chicken and rice dish its special flavor: little cans of fire-roasted green chilies from Hatch, New Mexico, my mother's favorite. The chilies were hand peeled, and sometimes you had to pick off a few specks of bitter, burned skin before chopping and adding them to your dish, but it was worth it. My mother was gone, but I had remembered some dishes whose aromas and tastes brought me tantalizingly and momentarily close to her. It was a rush of memory that, as I was sadly learning by making arroz con pollo over and over, was beginning to dry up with each attempt to tap into it.

It was Proust who most famously expressed his fascination and frustration with flavor and involuntary memory after drinking a spoonful of tea in which one of those small, scalloped-shaped sponge cakes, petites madeleines, had been dipped. "I drink a second mouthful, in which I find nothing more than in the first, then a third, which gives me rather less than the second. It is time to stop; the potion is losing its virtue."

When I was a child, I never much liked traditional New Mexican food. I grew up in the 1960s, and although both of my parents were Hispanic, we were raised mostly on a diet of iconic, processed American fare like Velveeta Cheese, Campbell's Cream of Celery soup, and my personal favorite, Welch's Grape Jelly. My sisters and I used this popular toast topper as flavored adhesive in the bacon sandwiches we made at our house every Sunday after Mass in the days when you had to fast all morning before Communion. These salty, sugary sandwiches (spread toast with grape jelly, lay on as much bacon as you can get your hands on, fold toast in half firmly until you hear the bacon crack) are perhaps best described as distant cousins of sweet and sour pork. They were always accompanied by fried eggs, and were invented by me, not my sisters, although they'll tell you differently. The few New Mexican specialties I did like, however, were of the sweet variety, such as Indian Fry Bread, served with powdered sugar or honey: a festive treat I first discovered at the annual New Mexico State Fair. If I had really wanted to impress Hughes I should have made sopaipillas, a similar New Mexican snack. Sopaipillas are hollow pillows of deep fried dough, the recipe for which is similar to that of flour tortillas, but instead of being flat they puff up as big as your fist when dropped in hot oil. I'm sure Hughes would have enjoyed the pleasant tactile ritual that goes with eating them: while the sopaipilla is still warm, tear off one of its pointy corners and pop it into your mouth—they're usually rectangular in shape, but my mother always made them triangular—and squeeze a bit of honey into the moist cavity from a plastic honey container shaped like a bear.

But I had not made sopaipillas, I had made salsa—and here in Brussels, where it was chilly enough that Jacky had made a fire in the fireplace, this typical Southwestern, undainty amuse-bouche not only looked unappetizing and out of place but, I realized as Jacky handed around glasses of white wine, should have been served with margaritas. Nevertheless, Hughes was eating the salsa with gusto, and wanted to know about me, that is, my food credentials. I told him I was a regular contributor to the weekend section of the *Wall Street Journal's* Europe edition, and loved writing about food and cooking.

To describe my life in culinary terms, however, would have been to conjure up a resume similar to that of many women of my age and background. I enjoyed baking when I was growing up; made faux tuna casseroles (canned tuna, sliced white bread, grated cheddar cheese) and the like in the toaster oven in my college dorm; learned how to cook economically "for one" by making dishes like lasagna, and eating the same thing, night after night, after getting my first job; became interested in cooking and dinner parties in the early years of my marriage after my husband and I moved to London where, inspired by one of the growing number of celebrity chefs who had begun to appear on British television, I once made Peking Duck by hanging it out to dry for hours in the open window of our mews house in South Kensington.

But I spared Hughes the details of my life in cooking, and found myself telling him instead about one of the most enjoyable articles I had ever written for the *Journal*. It was for a special weekend section devoted to sex and romance, and it was based on a potluck dinner I hosted to explore aphrodisiac food in Europe. The night of the dinner, friends arrived at our house, two-by-two, with dishes they had made with traditional aphrodisiacs from their home countries: everything from the phallic (eels cooked with tomatoes and mushrooms, from Jacky, who was Dutch/English), to the nipplelike (strawberries and raspberries with chocolate dipping sauce, from our Welsh friend, Melanie). In the questionnaire I asked the guests to fill out the next morning, most agreed that the "sexiest dish with the most arousing qualities" was *Pasta Aglio, Olio e Peperoncino,* cooked up on the spot by our Italian friends, a married couple who slipped away from the party into our kitchen to make the dish together. It was a simple spaghetti recipe made with olive oil, browned garlic and some tiny, powerful red chili peppers, *capsicum baccatum,* from Calabria—a nostalgic late-night snack from the couple's youth that young Italians might prepare after returning home from an evening out. (Aside from chili peppers, other foods considered aphrodisiacs in Italy, according to my friends, include balsamic vinegar, horseradish, and candied rose petals.) "My lips are tingling," someone murmured as we ate the spaghetti.

With the aid of some aphrodisiac cookbooks—the most eccentric and delightful of which was *Venus in the Kitchen,* published in 1952—I, too, contributed a few interesting dishes to the dinner, although I failed to source the main ingredient for one rather surprising antique recipe that had caught my eye called Testicles of Lamb.

I explained to Hughes that I hadn't been able to find lamb testicles anywhere in Brussels, even though it was not out of the question as I had once seen an enormous wild boar carcass lying on the sidewalk, freshly delivered, in front of a butcher shop, and because my local chain grocery store regularly sells beef tongues, and pigs' feet, ears, and tails. Eventually I telephoned the butcher at one of the largest, most exclusive gourmet shops in Brussels and he told me that it's illegal to sell lambs' testicles in Belgium, even though they're a delicacy in other countries.

"I have a whole book on testicles!" Hughes exclaimed.

Here was a man who knew everything and, as it would turn out, everybody.

Hughes was particularly delighted with the fresh coriander and the jalapeño in my salsa, but although we were eating from the same vessel, we were not experiencing the salsa in the same way. When he tasted it, he was reminded of and started talking about Thai cooking and other "exotic" cuisines that were growing in popularity in Belgium. The salsa I was eating, however, was laced with memories of the turquoise blue skies over Belen (Spanish for Bethlehem), the small town in New Mexico where my mother had grown up in an adobe house with eight siblings, and where my father, sisters, and I had eaten a traditional New Mexican lunch after visiting her grave a few days after her funeral.

Food, both the preparation and the eating of, takes up a lot of time in our lives. It's the backbone of each ordinary day, as well as a prominent feature of our celebrations, our religions, and our traditions. Food played a role in my mother's funeral, too—not just the offerings brought to the house, but in the form of family stories about cooking and eating. I told Hughes an anecdote shared with me by my cousin Elizabeth after the burial service as we looked in the direction of the church in which my parents had been married, and the elementary school next door my mother had attended. Elizabeth told me that our grandfather, during his lunch hour from work, used to run over to the school with sweet potatoes, hot from the oven, grown in his garden and baked by my grandmother for my mother and her siblings to eat. Perhaps the school didn't have a cafeteria or, more likely, my grandparents could not afford to pay for all of those school lunches.

Sometimes my mother prepared baked sweet potatoes for breakfast when I was growing up, just the way she had also eaten them for breakfast as a child: split in half lengthwise, gently mashed a bit, dotted with butter, sprinkled with brown sugar, and drenched in milk. On cold mornings, this steaming bowl of orange and red earth tones was milky sweet comfort by the spoonful. I told Hughes about this "recipe" of my mother's that I had forgotten about, and as I did I felt a rising sense of panic. I realized that as the taste of the arroz con pollo was gradually losing its power to grace me with a sense of her presence, so too would the sweet potatoes. Flavor-induced involuntary memory is a puzzling but universal experience, as intoxicating as it is fleeting, and I was terrified of losing it.

As I talked to Hughes, and we ate the salsa, an idea came

to me. I became fixated on understanding that ephemeral, mysterious thing called flavor. What if I put feelings aside for the moment and explored flavor thoroughly and logically? What does science tell us about flavor?

Flavor is an enormously complex subject influenced not only by our ability to taste and smell, but by sight, texture, temperature, sound, mood, ambiance, age, nationality, and gender, among other factors. It also involves a great many chemical, genetic, and other processes that occur in food as it's grown, preserved, or cooked. Focusing on the basics—how new flavors are developed and how flavor gets into food in the first place—I also wanted to explore my new obsession, flavor and memory. I needed to find the right scientists to talk to, and wondered how many women were involved in food science careers given that throughout history women all over the world have traditionally taken up the hugely important but often undervalued role of preparing millions upon millions of family meals. Indeed, I hoped my research would be as convivial as a shared meal, something like a particularly marvelous dinner my husband and I had had at the home of the Italian friends who had made the chili and garlic spaghetti for my aphrodisiacal evening. On the night of their dinner, they had invited us to come early and we all sat around the kitchen table stuffing homemade ravioli, assembly-line fashion, with savory pumpkin puree. If I was going to dissect and better understand the secrets of flavor, I wanted to learn by doing, tasting, and being surrounded by good company.

A week or so after meeting Hughes, we went out for lunch. It turned out to be a truly moveable feast involving several restaurants. The one he wanted to take me to was mysteriously closed, so we went to another across the street that he had wanted to review anyway. Here I watched him shift into restaurant critic mode, becoming increasingly overwrought—it was oddly chivalrous—as he pointed out to the waiter, the cook, and the manager that the tomatoes in our salads were bland. Hughes was as shocked and offended as if he had found the proverbial fly in the soup. He quickly ushered me out of that restaurant into another one he liked, but soon left in a huff, me trailing behind, because they wouldn't allow us to order just dessert. Finally we ended up in a fourth restaurant, talking about many things—his childhood in France, his memories of his grandmother's cooking and the evocative smell of her kitchen cupboards—as we enjoyed a tangy, aromatic dessert soup made with fresh oranges and whole cloves. We ate slowly. It was a slightly prickly but pleasant moment each time I found a clove in my mouth and sucked on it before placing it on the lip of the shallow soup bowl. And before we finished, Hughes promised to introduce me to some people who might be interesting for my research.

BRIGHT SWEETNESS

Let the progress of the meal be slow, for dinner is the last business of the day; and let the guests conduct themselves like travelers due to reach their destination together.

JEAN ANTHELME BRILLAT-SAVARIN
THE PHYSIOLOGY OF TASTE, 1825

Soon after, Hughes invited me on the spur-of-the-moment to what I thought was an ordinary wine tasting hosted by the Slow Food group in Brussels. Slow Food is a global, grassroots movement founded in Rome in 1989 that links the pleasure of good food with a commitment to local communities and the environment. It was a Sunday evening and I wasn't keen on going out, but in the name of research, I said yes. Before I left, I ate some yogurt and a piece of baguette—stomach liners, in case it was all wine and no food.

Belgium is a country where the notion of slow and excellent food comes naturally, fed by a rich, culinary split personality of Flemish and French influence. It's a microcosm of the historic passion for food in Europe, and the expertise and traditions that have evolved around it. A country where dining at home or in a restaurant is a sacred ritual that can last all evening, and where fresh bread is so important that some people have their daily loaf reserved at their favorite local bakery.

The event that night was at the École d'Arts Sasasa, a lively place that offers classes in cooking, art and dance, located in an Art Deco building in one of the city's residential neighborhoods. I arrived and was led by Hughes into the school's cooking demonstration room, where I joined about a dozen people standing around an oversized kitchen island. Everyone had that hushed, on-best-behavior look of children who've been promised a treat. To my surprise, small plates had been set out on the counter and something savory and aromatic was cooking in a covered saucepan on the stove. Hughes introduced me to the chef, Xavier Rennotte, a fresh-faced twenty-something who was busy with last-minute preparations. Xavier was well over six-feet tall, with closely cropped, slightly curly brown hair, and gray eyes. His smile was boyish, but his manner was quietly confident and exceptionally measured and mature.

The dinner unfolded slowly. I happened to be standing next to Xavier the whole evening, and we chatted a bit. I discovered he was a bank employee by day and a passionate beekeeper and trained chef in his free time. Xavier began the dinner by passing around pots of unusual honeys along with tiny coffee spoons. We took our time, sniffing and tasting, as he spoke in mellifluous French about the art of beekeeping and the exceptional and varied flavors and aromas of the different honeys. We transitioned gently into a four-course tasting dinner, and I quickly regretted the bread and yogurt I had eaten earlier. The dishes were Franco-Belge in origin, including fois gras and a selection of desserts from his fiancée's bakery. The highpoint of the meal was roast, marinated leg of lamb that Xavier carved and placed in small, warm ramekins that already contained mashed purple-hued, heirloom potatoes with a delicate chestnutlike consistency.

The star of the evening was a mysterious drink that had infused the lamb with a delicate flavor, various bottles of which also accompanied each course. Called hydromel (honey water) in French, it's known as mead in English. Along with wine and beer, mead is one of the oldest fermented drinks in the world, thought to date from around 8000 BCE—but instead of grapes or grains, its base is honey. It was this golden liquid that Xavier claimed was the only beverage aside from water that ever passed his lips. Mead had been virtually forgotten in Belgium, where hundreds of different types of beers are produced and hefty quantities of wine consumed, and Xavier was trying to rekindle an interest in this delightful old beverage one sip at a time. He served the mead to us in stemmed glasses, and it tasted like dessert wine.

Standing on the other side of me was a tall, slender, strikingly dressed young French woman named Sophie who, shy or aloof at first, had morphed into a chatty companion and had become my eating and drinking buddy for the evening.

"Cooking is my passion," she said at one point, as she

fussed over me and offered me second helpings. I had a feeling she found me gastronomically underprivileged, all because of my nationality. It wasn't the first time. When I first arrived in Brussels I once went to my local butcher, and after asking him what I thought were a few reasonable questions about an unfamiliar cut of meat, he looked as if he were going to follow me home to make sure I cooked it properly.

We made our way through glassfuls of mead, tasting different brands. Some of the meads were exceptionally flavorful, especially when paired with the food. Mead made me think of *Beowulf*, the Old English epic poem of heroes and dragons dating from the eighth century AD, awash with mead halls, mead cups, and mead drinking. Although mead is still fairly well-known in places like Britain, and is growing in popularity in parts of the US, I had never seen or tasted it growing up in America. So it was from reading *Beowulf* during college, I suppose, that I had formed a vivid but erroneous image of the drink as a frothy cross between beer and hard cider, served up in oversized beer steins. With each new mead we were served I took a sip, and then another, trying to pinpoint the flavor. It was hard to believe that this extraordinary beverage, which most of us were tasting for the first time, was nothing more than honey, water, and yeast.

In all, it was a wonderful dinner in which I had tasted a delightful new beverage, and had some lively conversations. After everyone helped clean up, people trailed out of the building mellow and satisfied, carrying doggie bags extraordinaire, but I left with more than just tasty leftovers, because Xavier had given me an idea.

"I'm developing my own brand of mead with a scientist from Louvain University," he had told me during the evening. The scientist was Professor Sonia Collin, an expert in brewing and honey research, and the project was in its very early stages.

From talking and listening to Xavier I understood quickly that his love of honey and mead were all-consuming obsessions, and that his mead project was as maniacal, and taste/memory-driven, as it was gastronomically obscure. Indeed, Hughes had found just the right contact to start me off. A few days later I found my old paperback copy of *Beowulf*, flipped through it searching for ancient descriptions of what I had recently tasted, and saw mead described as "bright sweetness." A fitting description for the evening I had just had, as well as for the elusive mead flavor Xavier was hunting for.

THE SINGING BAGUETTES

The little country of Belgium, which has been through so many invasions and occupations by armies from other countries that it is hard to keep count, has a proverb that reflects its deep agrarian roots and pragmatic, country-folk wisdom: 'As long as there is bread on the table, there is hope.'

RUTH VAN WAEREBEEK,
EVERYBODY EATS WELL IN BELGIUM COOKBOOK, 1996

A passion for food runs deep in Xavier's family, as well as in the family he was about to marry into. His grandfather had been a tea importer who, perhaps like Xavier, was a pioneer in his own day, as Belgium is a country where coffee is more popular. Xavier was engaged to Claire Wyns, co-owner, manager and granddaughter of the original owner of Au Vatel, one of the oldest boulangerie-pâtisseries in Brussels.

Xavier and I next met up at Au Vatel, named after François Vatel, the seventeenth-century French chef who invented Chantilly cream—beaten cream and sugar (whipped cream)—for an extravagant banquet given to honor Louis XIV at the Château de Chantilly. The story goes however, that Vatel did not live to savor his triumph; he committed suicide, it is said, perhaps apocryphally, because the fish delivery was late. Au Vatel sits in Place Jourdan, an old city square dotted with restaurants, including a popular pizza place called Mama Roma's where toppings include thinly sliced potato cooked in cream—unexpectedly delightful and in perfect harmony with the thick, airy pizza crust. Another notable place, located on the island at the center of the square, is the Friterie Chez Antoine—one of many such *frites* shacks in the city—renowned for its chunky potato slices, never referred to as French Fries here, deep fried in beef fat and served up hot and salty in a paper cone. Ketchup may be requested, but Belgians prefer to dip their *frites* in mayonnaise.

I arrived early at Au Vatel and ordered a *lait russe,* a richer, Belgian version of café au lait, which was served along with a tiny marzipan morsel. It was Armistice Day, a public holiday, and people were in a festive mood and, as usual, made any excuse for getting together to eat. Xavier showed up, ordered a bottle of sparkling water, and spoke about his childhood, his mead-making plans, and his dream of keeping bees atop the grassy roof of one of the largest water tanks in Brussels, the size of a football field, which he had spied from the window of Claire's apartment above the bakery. In order to be licensed to keep bees in Brussels, the hives have to be kept a good distance away from the general public, and in a city like Brussels, a good place to go is up.

A beekeeper since he was a teenager, Xavier also dated

his love of mead from about the same time. At fourteen, he had his first sip when his family visited an elderly man near their home who made mead flavored with elderberries. At seventeen, as a member of a Belgian Boy Scout troop visiting Quebec, he discovered the taste of mead once again, this time flavored with blueberries.

From a young age Xavier had a singular vision that he would carry with him into adulthood. "I saw honey flowing, like gold," he said.

Now twenty-seven, and with a grant from the Belgian government, he had hired Sonia Collin to help him give mead a taste makeover. The project was perfect for Professor Collin, an expert in brewing and in honey at the Université Catholique de Louvain. But why, I wondered, did Xavier need to work with a scientist to help him give a new taste twist to an old beverage? He explained that it was his godfather who had given him some sage advice before he began the project. "Learn from the beginning the scientific way," his godfather had said. "The best way to understand something is to go deep inside it."

Eventually, Xavier invited me upstairs to Claire's apartment to conclude our conversation with a glass of elderberry cream, a strong liqueur sometimes used to make champagne cocktails. He wanted me to taste it so I could connect, however remotely, with the memory of his first drink of elderberry-flavored mead. The charming equivalent, I suppose, of Proust handing you a cup of tea and a slice of sponge cake in the hope that it would give you a snapshot of his madeleine-filled childhood. To get to the apartment we had to walk through a door at the back of the café, where we paused on the landing of the stairwell to inhale the scent of warm, sugary butter from the pastry kitchen below, and then up a few stairs to the hanger-like room where the bread is baked.

Xavier exchanged a few words with the baker on duty.

"We'll come back so you can hear the baguettes sing," Xavier said to me mysteriously, and we moved on.

Up in Claire's lovely old apartment, with its high ceilings, Xavier poured each of us a small glass of the forty-proof elderberry liqueur, and we drank it standing in the kitchen. The liqueur was delicious, and for a second I really was able to imagine him, a young teenager, taking his first drink of mead.

Xavier began to explain how demanding and time consuming it was to create a new and specific mead flavor. The primary issues he and Sonia were working on were how to retain the aroma of honey in the final product; whether it would be possible to make unwanted flavors disappear by changing the speed of the fermentation; and how to create a product without sulfites.

I asked Xavier if he could put into words the type of flavor he was trying to create, but he said he would only know it when he tasted it. It was obvious from his expression that this frustrated him greatly; that he would have gladly explained it to me, to Sonia, to everyone, if he could have.

The answer was all rolled up in the honey-infused tastes, sounds, sights, and smells of his youth. Xavier was taking that universal, electric sensation—memories sparked by a certain taste of something—and trying to capture it in a bottle.

We put down our glasses and he led me back down to the baking room where tall metal racks holding dozens of exquisitely long, slender baguettes had just been wheeled from the ovens. We stood there, warmed by the heat as if standing in front of a row of pleasantly yeasty radiators, and it was just as he had promised, the baguettes *were* singing. It was a curiously comic shrilling, like a chorus of whistling teakettles. Just as unexpected and funny as the first time, when I suppose I was four or five, I remember standing in front of the stove and hearing the rat-a-tat-tat sound of popcorn ricocheting inside a covered metal pan.

Xavier gave me a moment to drink it all in, this singular combination of heat, aroma, and sound, before he pointed to the hairline cracks on the loaves' golden crusts, from which the steam that had built up deep within the loaves was now escaping.

Yeast had made the bread rise, but it wasn't until I returned home that I discovered that it was also responsible for making the baguettes sing. In my treasured copy of *McGee on Food & Cooking*—the bible of food lovers everywhere—I learned that yeasts metabolize sugars for energy, producing carbon dioxide gas and alcohol as by-products.

In making beer and wine, carbon dioxide escapes, and alcohol remains. But in bread, both are expelled by the heat of baking.

By chance, in an old Brussels bakery, I had just had the most melodic and aromatic introduction to food science anyone could have wished for. Without yeast there would be no bread and wine, no cakes and ale, and no Au Vatel or Xavier's dream of a perfect hydromel.

LIBRARY OF YEAST

When the dog-days set in, take some spring water; to three parts of water add one part of clarified honey; put this mixture in earthenware vessels and have it stirred by your slaves for a long time. Leave it out in the open, covered with cloth, for forty days and forty nights.

NORMAN DOUGLAS (ED.),
VENUS IN THE KITCHEN, 1952

After making an appointment with Sonia, I boarded the student shuttle for the Université Catholique de Louvain located just south of Brussels. The university is known for its scientific research, and was the place where, in 1996, under the supervision of Professor André Goffeau, one hundred European laboratories published the complete sequence of the yeast genome.

The Louvain campus is a maze of concrete paths and modern buildings. And inside one of these contemporary buildings is where the Department of Brewing and Food Industry conducts research involving foodstuffs with ancient roots, such as bread, honey, olives, wine, and beer—and now, for the first time, mead. Sonia Collin, the head of the department, ran a research lab with some unique features: small, plastic beer barrels were stacked in one of the hallways; two white-coated scientists rushed past me, test tubes in hand, out of a room labeled "chocolate lab"; and on a door in one of the corridors there was a warning sign, the English half of which had gained something oddly canine in translation: SNIFFING, *Ne déranger pas* (do not disturb).

I met first with Sonia's lab technician, thirty-one-year-old Aurore Timmermans, who asked if I would like to see the department's library of yeast, which was her responsibility. She led me down the hall to a room no bigger than a walk-in closet, with a locked door. Inside was a refrigerator holding some two hundred test tubes filled with precious yeast cultures, used primarily for brewing. Breweries in Belgium buy yeast from the department, made up for them in big batches, but they also bank their own signature yeasts there for safekeeping. To keep the yeasts vigorous Aurore replaced each tube with a new batch of yeast every six months. The test tubes lined up in the refrigerator made me think, not of a library, but of a sperm bank for the procreation of an endless lineage of Belgian beers.

Aurore took me back to the meeting room where we had begun our conversation, and soon after she left, Sonia rushed in. She was a petite woman in her forties, her long, blond hair tied up in a ponytail. She didn't have much time, but took a moment to make an espresso for me, and out of her pocket she pulled a sweet amaretto biscuit wrapped in pink paper to go with it.

Although she works with honey, she doesn't eat much of it or other sweet foods. But she is intrigued by a link between the lab and the dinner plate, and does enjoy a bit of honey in *salades de chèvre chaud en croûte,* warm goat's cheese wrapped in pastry, served atop a green salad, a common luncheon dish or appetizer in Belgian restaurants. But as for honey in the lab, its pleasures are sensual and mysterious in other ways than taste. "It's an exceptional source of aromas," she told me. "Certain molecules in honey are exactly the same as certain molecules in Sauternes wine."

The day I visited the department was an auspicious one in the life of Xavier's project, for later that afternoon she would give him the short-list of the six yeasts she had chosen for them to work with. Xavier later showed the list to me, and I thought it read like a repetitive, boozy rendition of "The Twelve Days of Christmas": *Two French wine yeasts, two Spanish wine yeasts, one common mead yeast, and one Champagne yeast.*

Although meads traditionally have been, and still are, made with the addition of fruit and spices, Xavier's goal for the moment was to use only the three basic ingredients, honey, water, and yeast, with no additives, thus showcasing the aromas and flavors of the honey. Xavier had already chosen some standard honeys for the project: orange blossom, acacia, and a multifloral honey made from French and Belgian wildflowers. But in the future, he said, he would like to try thyme honey, although it's rare and expensive. He was also curious about buckwheat honey, which is dark and strong, but perhaps the right yeast and a bit of aging could tone it down, he thought.

During that meeting, Sonia and Xavier also decided that they would like to have their first drinkable product in about six months.

"Then we'll put it to sleep in a *cave* and let it age," Xavier told me. I loved the way Xavier's English was often infused with sweet images from his native French.

But before the mead could be put to sleep, there would be months of testing and calculating and sniffing in the lab. Xavier and I planned to meet regularly so that he could keep me updated. I was impressed that his mead project was in such good hands, and amazed to have discovered a secret world where so many people were tweaking, shepherding, and researching traditional foodstuffs for our enjoyment. Seeing so many types of yeast so carefully nurtured also made me curious about beer, another ancient beverage. Xavier's mead project was my starting point, and my home base in the coming months would be Au Vatel. But I also wanted to discover who else in Belgium was quietly obsessed with the science of flavor.

BEACH PARTY IN A BOTTLE

Is not air the most absolutely and most immediately indispensable element to us, and is not smell truly the only sense that perceives some parts of it? Scents, jewels of the air we need to breathe, do not embellish it without reason.

MAURICE MAETERLINCK,
THE INTELLIGENCE OF FLOWERS, 1907

Producing a new type of mead in Belgium was unique. There was plenty of beer making around, however, as beer is a popular drink in the country and used in many traditional dishes such as Flemish Beef Stew. Belgium, a small country, hosts some 125 breweries that produce a wide range of beer types and flavors. The country is also home to six of the seven breweries in the world still located in Trappist monasteries, which house an order of contemplative monks. (The seventh Trappist brewery is next door, in the Netherlands.)

The production of beer is traditionally a man's domain, with a few exceptions, one of them being An De Ryck, owner of the Brouwerij De Ryck, a small, independent, award-winning family brewery located west of Brussels in the Flemish, or Dutch-speaking part of Belgium. Early one

Saturday morning, I took a train to the De Ryck Brewery, which was begun in 1886 by Gustaaf De Ryck, An's great-grandfather, on a former farm in the town of Herzele. The brewery has been in operation continuously (except for a brief period during World War I), passed down to Gustaaf's granddaughters, and now to An, who, in turn, has a daughter in her twenties involved in the commercial side of the business.

The first question I had about brewing was whether it was a craft, a science, or both?

"Both," said An. "You can't be a brewer without knowing about fermentation and other processes."

She had a degree in brewing engineering from KaHo Sint-Lieven, a university college in the Belgian city of Ghent, where she had studied chemistry, biochemistry, and microbiology, along with brewery technology. At the moment, the flavor of beer was a commercial as well as a scientific concern of hers, she said. Belgians were drinking less beer each year, in part because of the country's increasingly strict drinking and driving laws, but also because young people were drinking more wine and wine coolers, or alcopops as they're called in Europe. The result was that exports were becoming more important to the breweries. But exported beer sits in the bottle longer, and that can be a problem.

"The enemy of beer freshness is oxygen—and we want to know what we can do during production to avoid the influence of oxygen," said An. De Ryck belonged to a consortium of thirty breweries working together with two universities to study beer. Their current project was to understand what happens to the flavor of beer when it's stored. She gave me copies of two recent scientific articles on beer flavor by scientists at one of the universities, KU Leuven: "Contribution of staling compounds to the aged flavour of lager beer by studying their flavour thresholds" and "Optimization of a complete method for the analysis of volatiles involved in the flavour stability of beer by solid-phase micro-extraction in combination with gas chromatography and mass spectrometry." This was a lot for me to take in on a Saturday morning. I did my best to keep up, but I felt a certain sinking feeling, worried that An didn't realize I was there to learn from scratch, not to match scientific wits with her. I figured I had two choices: quit now, or attempt to fake my way through a question and answer session with her.

Instead, I chose a third option: I asked for a drink. If we were going to discuss the flavors of beer, then I needed to taste one. An and I settled down in the shop area of the brewery where customers could enjoy a glass of beer on tap seated at tables made of old beer barrels, and she poured me a glass of one of her signature beers, a fruity brew called Jules de Bananes.

Different types of beers, like French nouns, can be masculine or feminine, although the generic French word, *la bière*, is feminine. The fruity ones, like Jules, are usually looked upon as girly drinks. Jules was born in 2005 as a flavor trump card in hopes of increasing beer sales among Belgian women, who seem to prefer beers with the taste of cherry or raspberry. So, thought An, why not give them a banana-flavored beer, perfect for a beach party or a sunny sidewalk café? That's how Jules was marketed by the brewery, using the power of suggestion to attempt to manipulate flavor-induced fantasies. It was a sunny fantasy I was happy to embrace in the chilly brewery shop, which was so cold I had to keep my coat on. I tasted. The beer was sweet, but its aftertaste was pleasantly bitter. Although it tasted like banana, it still tasted like beer, and the unfamiliar pairing of the two flavors was, to my mind, a brilliant combo. Banana flavored beer, it seemed, wasn't that outrageous, when An told me of other brewers who made beers with flavors like chocolate, coconut, and tobacco.

I was about to drink again, when suddenly An took the small, stemmed beer glass from my hand, cupped it in both her hands, swirled, tilted her face close to the glass with one abrupt movement, and sniffed. She was like a stage mother fussing and fretting over her offspring until the child performed perfectly for the audience. The look on An's face lay somewhere between mild letdown and stony seriousness. She swirled, coaxed, and sniffed again, willing Jules to give up the full range of its volatile molecules.

She returned the glass to me and said, "The aromas will increase as the liquid warms up in your hands." She said it in a tone that was both resigned and determined. She was at the mercy of time (would I wait until it warmed up properly?) and temperature (did I have enough heat in my hands to warm Jules?). She encouraged me to slow down, to enjoy the aromas as well as the taste. And she was right. As I sniffed and sipped, its sweet, complex flavors became more pronounced.

An also sold a variety of food products made with De Ryck beers: hazelnut cheese, paté, chocolates, marmalade, and sorbet. Like Xavier, she gambled that the flavors she liked would be appreciated by others. Flavor development is not only about ingredients and preparation methods, but the perceptions of the people doing the tasting. An believes women are better than men for the job. Not only are they better at explaining it and at comparing differences, she said, but age counts as well. "At fourteen, girls' ability to taste is at its peak," she said.

After I finished my beer, An and I left the shop and went across the cobblestone courtyard to the brewery, still in the original old farm building. The minute we entered, I felt the difference in the air, heavy and humid and, as I vividly imagined it, crawling with yeast. De Ryck uses its own yeast, cultivated on site, named after the brewery, and unique to the area. A sample of this yeast, kept dormant in liquid nitrogen, is banked at KaHo Sint-Lieven University.

"We isolated the yeast in the 1970s, and it's probably the same yeast that my great-grandfather was working with over one hundred years ago," said An.

The brewery, on the day I saw it, was at rest. The building felt as somber, dark, and ghost-filled as it might have been when production stopped during World War I after German troops confiscated the copper brewing kettles, most likely to be melted down and recycled as munitions. On the ground floor, An showed me the large copper filtration barrel, which filters the wort, a clear sugary solution that is produced after barley has been germinated, heated, milled and soaked in hot water. We walked up to the next floor and entered a dark room that looked like it had once been a hayloft. An reminded me to close the old, heavy doors carefully to keep mice out. Then she invited me to stick my head into the top of the malt mill, where malted barley is crushed. Malt is nothing more than barley grain that's gone through two steps: soaking it in water to encourage germination; and kilning, to dry it at varying temperatures over a period of anywhere from sixteen to sixty hours, depending on the flavor and color of the malt being made.

I inhaled the lovely, nutty smell of the malt, but was unprepared for the rush of memory it sparked off. I told An that the aroma from the malt mill reminded me of chocolate-covered malted milk balls, a favorite childhood candy with a crunchy honeycombed center. I remembered buying them once during a family trip to the local general store in search of school supplies for the first day of elementary school. Part of the ritual was something all children I knew did, but I can't imagine anyone doing today: I asked the man behind the tobacco counter for an empty cigar box to be used as an aromatic desk caddy for my newly purchased crayons, pencils, scissors, and glue. Oh the smell of crayons after they've been stored in a cigar box! My sisters and I ate the candy on the way home—I liked letting the chocolate dissolve in my mouth first, before crunching into the malt—as we wriggled around in the backseat of the car (there were no seatbelts in the back in those days) listening to Andy Williams sing "Moon River" on the car radio. Our favorite foods from childhood are accompanied by colorful back stories, are as personal as a fingerprint, and are introduced to us at a time when we are as impressionable as bread dough.

My mentioning the malted milk balls prompted An to describe her own childhood, a storybook life that seems anachronistic today. She was an only child, and grew up playing in the brewery until she was around twelve when she was given little jobs to do. Gradually she realized she wanted to get beyond her family-style apprenticeship, and master the job by learning about the science behind it all. "There comes a point when you're curious to know more about the work," she said, and eventually she ended up studying at the same university her father had attended.

Everything in An's family and business worlds is bundled together in the brewery compound. To get to the home An shares with her husband, one merely has to walk behind the cash register in the shop and open a door. Her house, in turn, is connected to her parents' house next door. Across the courtyard is the brewery, which is connected to a building that holds her husband's pharmacy and her small laboratory. Inside the lab's fridge was the De Ryck yeast, sitting in test tubes, as well as the banana flavor she uses for Jules, which had been made into an easy-to-work-with liquid form at a so-called flavor house. She regularly puts the yeast under a microscope to test it for infection by bacteria. Although some Belgian beers are produced through spontaneous fermentation—exposure to airborne wild yeast and bacteria—her type of beer would be ruined if bacteria managed to get into it.

"Bacteria could make the beer sour or give it another taste. Sour is good in other beers, but not ours," she said.

After our tour and tasting, An drove me back to the train station, where I rattled and clinked as I staggered toward the platform with both my handbag and a plastic bag filled with beer bottles. I boarded the train with regret that I had only skimmed the surface of the science, the history, and the drama that had passed through that little brewery for nearly 125 years. But there was a consolation prize: Jules was coming home with me. Liquid proof that if An could bottle the idea of fun in the sun, Xavier might also succeed in developing a lively flavor for his mead.

LEARNING WHILE INTOXICATED

For, after all, it is housewifery to which nearly all the arts and sciences bring their secrets. Home and comfort, food and drink—it will be a long time before we can get quite away from the need of these things. To introduce science and order into the domestic kingdom is a task worthy of the finest intellect.

MRS. CHARLES ROUNDELL
AND HARRY ROBERTS,
THE STILL-ROOM, 1903

Talking to Xavier and Sonia about mead making was basically straightforward, as they were following in prehistoric man's footsteps by allowing water, yeast, and honey to ferment. But fine-tuning the flavor in order to recreate Xavier's childhood memories seemed as delicate, skilled, and near-impossible a process as the work of a Japanese scientist I once read about who had made violin strings out of spider silk. Beer making, on the other hand, was robust and familiar, and had long been associated with the home.

In beer making, as in cooking, the taste of the final product is influenced by variations in both the type and amount of ingredients. Part of the special flavor and aroma of the De Ryck beers came from Belgian malt made from French barley, along with hops from all over the world. But that's just the beginning. I did some reading and learned that some of the flavor also arises from the fermentation itself.

I contacted Debbie Parker, flavor training manager at Campden Brewing Research International (BRI), in England, a research and development organization for the food and drink industry, to learn more. Debbie's job was to teach people in the industry more about the product they were developing and selling by learning how to smell and taste it accurately. She had a degree in biochemistry, a doctorate in brewing science, and a post-grad certificate in sensory science. She had also been a professional beer taster for eighteen years, and along with her husband, owned a pub called The Miners Arms in the village of Dunton Green in Kent.

When I arrived at BRI, located in an old manor house in the picturesque countryside of Surrey, Debbie took me into the sensory facility area where we sat at a table next to a small bar with some hefty beer pump handles. Debbie had long red hair, a big smile, and a lot of enthusiasm. On the table, she had several books she thought I might be interested in.

From *Beer: The History of the Pint,* by Martyn Cornell, I learned that brewing was originally a female occupation, with women known as brewsters or alewives taking turns brewing for the local community in their homes, incorporating beer making into their other domestic duties. I had brought some books along with me, too, including a small pea-green hardback Debbie liked but had never seen before, called *The Still-Room* by Mrs. Charles Roundell and Harry Roberts, published in 1903. In the chapter on home-brewed beer the authors say that tea drinking was once seen by some as "a destroyer of health" compared to the benefits of beer. "How different . . . is the fate of that man who has made his wife brew beer instead of making tea!"

I asked Debbie a few questions, hoping to clarify some of the material I'd been reading. But it was like pulling on a thread of a sweater, and watching the whole thing unravel. My first question was about the generation of flavor that continues after the death of the yeast cell.

Debbie's tantalizing answer included whole concepts I didn't understand alongside familiar, comforting words like "butterscotch flavors." Like the two research papers that An had shown me, this was high-octane flavor science, but I needed a simple explanation, and Debbie was happy to summarize: beer has around one thousand different flavor compounds which may spring from hops, malt, and the yeast—although some 60 percent of a beer's flavor comes from the yeast.

Next we began my beer sensory lesson in the narrow kitchen next to the pub room. Before me on the counter were four large glass jars filled with malt kernels. The malt graduated from pale to dark brown, and the jars were labeled "lager malt," "ale malt," "crystal malt" and "chocolate malt." Especially tempting was the "chocolate malt." One by one, Debbie let me taste a single sample, a grain of germinated and kiln-dried barley from each jar. The lager malt tasted nutty and sweet; the ale malt was richer, like sweet cashew; the crystal malt tasted bitter. Despite the fact that each one in turn was becoming less tasty, surely the last would taste like chocolate-covered malted milk balls, I thought. But quite the opposite—the "chocolate malt" tasted as acrid and unpleasant as a charred coffee bean. I gathered by the look on Debbie's face that I wasn't the first one to be dismayed by the confusion between the name and the taste.

"The higher the temperature and the longer the grain stays in the kiln, the more flavor and the more color it gets," Debbie said. Chocolate malt has a lot in common with black pepper: unpalatable on its own, but when added to food (or beer, in this case) it can be an indispensable flavor and color enhancer.

We moved from the kitchen to a room next door, which had a row of little cubicles and some computers. Here Debbie gave me a selection of little "sniff pots" on which to test myself while she provided a running commentary on brewing, aided by a power point presentation.

The flavors I was sniffing had been purchased at a so-called flavor house—like the one where An had purchased the banana flavor for Jules—and were used in training brewing professionals to build up a "flavor profile" of their product. This helps them become skilled in accurately scoring both the types and intensity of the flavors in their beer, in order to make sure these flavors didn't drift over time.

I sniffed all the little pots, including three containing the aromas of different essential oils of hops. One was citrusy, one smelled like cloves, and one smelled grassy. In another book I had brought with me, *Brewing,* by Ian Hornsey, published by the Royal Society of Chemistry, I had read that the hop is related to the stinging nettle, bog myrtle, and hemp. The hop most commonly used in brewing, *Humulus lupulus,* is a climbing perennial herb. For the brewer, the useful part of the plant is the female cone, which contains all of the resins and essential oils that give beer both its bitterness and some of its aroma.

We returned to the pub room for a BYOB beer tasting. I had carried into England six different Belgian beers that I had bought at my local grocery store on the spur of the moment the night before—in truth, selected on the basis of their pretty labels, and hopefully interesting flavors. Though I had chosen at random, I was richly rewarded, for each represented a different beer type. They included a spontaneously fermented peach-flavored lambic beer, and a pale blond beer that was triple fermented with three different yeasts (that I had chosen because of the pink elephants on its label).

We began with a St. Feuillien Brune/Bruin, a top fermenting ale in which the yeast rises to the surface during fermentation. It came in a stout, curvy little brown bottle with an attractive label from a sixteenth-century painting showing a jaunty pilgrim in pink pantaloons carrying a basket of hops on his back and striding toward the ancient Abbaye St-Feuillien du Roeulx in Belgium.

Debbie poured two glasses of the beer, and immediately placed a plastic cover over each, so the volatile molecules wouldn't escape. I did everything she did. She swirled, I swirled. She held her glass up to the light; I held my glass up to the light.

She described the beer as reddish-amber in color, topped with a cream-colored head. The rich, dark color of the beer, she knew from just looking, was a product of chocolate and crystal malt. We each took the lid off our glass.

"What does it smell like to you?" she asked.

I sniffed. I pondered.

"Beer," I said.

She waited.

"Belgian beer," I said, trying to be more specific.

We tasted, and out of a few sips came a torrent of descriptive words from her: "Spicy, phenol, smoky, malt, toffee, caramel, banana; in all, it's a well-balanced beer—not too bitter, with good body."

She sipped again.

"It tastes like those old-fashioned, banana toffee chews we used to eat as kids," she added.

I admired her ability to pinpoint all of the flavors in the beer, and noted with pleasure that she had neatly summed up the overall flavor by comparing it to a taste from childhood.

We walked over to the main building where I met her boss, Caroline Walker, the newly appointed director of brewing at the Nutfield site. Caroline had a doctorate in biochemistry from the University of Bristol, and had worked in Denmark as a fellow at the Carlsberg laboratory of the Carlsberg Brewery in Copenhagen. The laboratory was established in 1875 to develop a complete scientific understanding of the malting, brewing and fermentation processes involved in beer making.

We picked up on the earlier conversation I'd had with Debbie about women making beer at home, for domestic use, centuries ago. "I have a suspicion that women used herbs in beer to medicate the family," Caroline suggested. It would have been a delicate process, as any mother who has tried to increase vegetable intake in the family diet knows: the final product would still have to taste good in order for the nutrition to go down. It was interesting to think that women in the past made beer in the home without knowing the names or workings of the complex processes they were dealing with, and now it was all being explained to me by two women who were experts in brewing science.

We walked over to the company library where Caroline found a book she thought I might be interested in: *Sacred and Herbal Healing Beers: The Secrets of Ancient Fermentation*, which included recipes for mead as well as beer. One delightful sounding recipe, for Heather Mead, called for six pounds of heather honey, ten cups of light pressed (unwashed) flowering heather tops, four gallons of water, and yeast.

I left this cozy, bucolic brewing institute with the flavors of both beer and mead on my mind. Despite their differences, mead and beer are a lot like kissing cousins. They are mentioned side by side in the *Kalevala,* a nineteenth-century Finnish epic poem, and have also shared honey as a common ingredient. In the *Kalevala,* a bee is sent to collect honey for beer brewed for a wedding feast. Honey is also mentioned in the "Hymn to Ninkasi," an oft quoted Sumerian beer "recipe," or drinking song, inscribed on clay tablets in the eighteenth century BCE. More recently, after President Barack Obama showed an interest in home brewing, honey from the White House beehives was used to produce three types of honey beer: an ale, a porter, and a blonde. Sam Kass, White House assistant chef and the senior policy advisor for healthy food initiatives, wrote on the White House blog: "Like many home brewers who add secret ingredients to make their beer unique, all of our brews have honey that we tapped from the first ever beehive on the South Lawn. The honey gives the beer a rich aroma and a nice finish but it doesn't sweeten it."

LIQUID COURAGE

. . . liquids require receptacles. This is the great problem of packaging, which every experienced chemist knows: and it was well known to God Almighty, who solved it brilliantly, as he is wont to, with cellular membranes, eggshells, the multiple peel of oranges, and our own skins, because after all we too are liquids.

PRIMO LEVI, *THE PERIODIC TABLE*, 1975

Soon after my return home from England, I met Xavier at Au Vatel on a chilly spring day. It was weeks before Easter, but all of the bakeries and pâtisseries around town, including Au Vatel, were displaying handmade Easter chocolates made from fine Belgian chocolate, beautifully packaged in cellophane and curly ribbon. There were pastel-colored eggs lying in dark, bittersweet chocolate nests, ornate white chocolate church bells, and my favorite: old-fashioned storybook animals, such as milk chocolate rabbits driving milk chocolate motor cars.

It was late morning, and at that time of day Au Vatel was peaceful. Customers were drinking coffee alone or chatting quietly with someone in the lull before the lunch hour. Others came in, scanned the freshly baked loaves of bread on the wooden shelves behind the counter, made their choice—a crusty baguette, some buttery croissants, or a *cramique,* a brioche-type raisin bread with a glossy bronze crust—and left with their purchases wrapped in a brown paper bag.

I found Xavier at a table at the back of the bakery, reading a newspaper article about Colony Collapse Disorder (CCD), a mysterious and deadly bee colony phenomenon that in recent years has been reported widely throughout Europe and North America. He spoke briefly about CCD before turning the conversation to his mead project. Sonia had produced a batch of mead that succeeded in retaining the honey aromas. The three types of honey they had used were a secret, but he did tell me they were from Sicily, the South of France, and Hungary.

This first batch of mead was pleasant, but not perfect. "For me, it tasted a little too watery," said Xavier. "It wasn't complex enough. It doesn't have the length that good wines have, and it didn't feel right in the mouth." He thought adding fruit might answer the problem. He gave the impression that he was still a ways off from his ideal mead, but that he was happy to have reached this stage. Along with improving the flavor, Xavier was urging Sonia to reduce the sulfites so his mead would be as natural as possible.

Xavier had just bought part of his parents' land in the Belgian countryside on which he planned to build a meadery. He was delighted to have discovered water under the land and his plans were to dig one hundred meters down to get at it and use it as one of his mead ingredients. To test the water, he had recruited a scientist from the Institut Meurice, a research institute in Brussels. She was still working on it but had already found one defect.

"There's too much iron in the water," Xavier said, which gave it a metallic taste.

I sensed Xavier had reached a certain momentum in his project where new possibilities and sub-projects were occurring to him at hyperspeed. This was critical for exploring new ideas but, equally, he would have to know which to focus on and which to abandon. Not always easy with a monomania like his: honey-colored glasses through which he saw almost everything, even the survival of the species: "If bees die, we die," he had said when we were talking about CCD. Honey and mead were everything to him: life and death; hearth and home; science and business.

Xavier's plans to install his hives on the water tower he could see from Claire's apartment had fallen through, so he decided instead to install three hives, the maximum number allowed without a permit, on the roof of the bakery. He planned to sell the honey in the bakery itself, but would not be using his own honey in his mead because city honey has too many variable flavors and aromas. Unlike, say, lavender honey from the countryside, city honey is feral by comparison, made from a huge and ever-changing variety of different flowers and blossoms found in parks, window boxes, and private gardens.

Xavier was also thinking about packaging. Every liquid food needs a container, and I had assumed he would only be putting his mead in wine bottles and his honey in ordinary glass jars. But he also had plans to encase drops of each in a solid edible skin, probably made from agar, an algae derivative. Though firm at room temperature, these drops would melt when spooned into a hot liquid—tea, for example—leaving only the flavor of honey or mead.

Agar, or agar-agar, as it's also known—an echo-like reminder of the way it has bounced back-and-forth between kitchen and lab for over one hundred years—was an interesting choice. I had recently seen it listed as an ingredient in a Czechoslovakian World War II cookbook in which it was described as "a form of dried seaweed with setting properties," useful for making cakes, small pastries, and fish dishes. But the first time many people, myself included, remember ever coming across agar was during high school where it was used to grow bacteria in Petri dishes. Indeed, it has been a fixture in microbiology since the late nineteenth century, having arrived in the lab with the help of some female lateral thinking. Around 1881, Fannie Hesse, the wife and laboratory assistant of a German physician and scientist, Walther Hesse, suggested to him that agar might be useful as a medium for growing bacteria in the lab. Fannie used it in her jellies and puddings, which stayed solid even in summer, and had learned about it from her mother in New York, who learned about it from friends who had lived in Asia. Walther Hesse found agar's thermal properties far better than gelatin, and suggested it to his colleague Robert Koch (who had also tried potato slices, which weren't ideal). Koch, a German physician considered one of the founders of bacteriology, embraced the new medium, and the Petri dish, used to hold the agar, would later be developed in his lab.

Agar's latest popular reincarnation has brought it back to the kitchen. I had recently seen a demonstration of molecular gastronomy at the same institute in which Xavier was having his water checked out, and had been offered something that looked like golden fish eggs—and in fact had been referred to as caviar, but were really tiny agar balls filled with a meaty, savory liquid.

Xavier's idea to encapsulate a certain burst of mead or honey flavor in agar reminded me of how nature sometimes packages liquid food. An orange, for example, is just a juice box that doesn't need a straw: you can peel it, and never spill more than a few drops of its liquid, each crescent-shaped section holding the juice in place. Cut the orange, or take a bite out of one of the sections, and most of the juice still doesn't spill out because it's further packaged into smaller juice sacs.

After leaving the bakery, I was mulling over the commercial potential of honey-and-mead-filled agar balls, and the intricate engineering of orange membranes, when suddenly a vivid image popped into my head: Hieronymus Bosch's phantasmagorical fruit-laden central panel of his famous triptych, *The Garden of Earthly Delights*. I found a reproduction of the painting in an art book we had at home, and it was as I remembered: an aphrodisiacal party to end all aphrodisiacal parties. At the bottom I saw a peaceful, dreamy scene. There was an oversized pinkish fruit pod, out of which tumbled dozens of deep elderberry-purple, caviarlike berries. And slumped over the fruit pod was a sleepy-looking fellow who looked as if he had just fallen into a lovely trance, having perhaps just eaten some of the berries and, behold, had discovered the aroma and taste of their juice was good; perhaps reminiscent of honey water.

I was rooting for Xavier.

II. HIDING AND SEEKING FLAVOR

LICORICE ON THE INSIDE

But, as a general rule, we are bound to advise all mothers to abstain from such articles as pickles, fruits, cucumbers, and all acid and slowly digestible foods, unless they wish for restless nights and crying infants.

ISABELLA BEETON, *MRS. BEETON'S HOUSEHOLD MANAGEMENT*, 1861

Flavor is like oxygen: because we have endless supplies of it, we sometimes take it for granted. Without oxygen, we would die; without the ability to smell and taste, we might also end up in a bad state by losing our desire to eat, or by eating poisonous or spoiled food. But where does our lifelong companionable relationship with flavor begin? To find the answer, I flew to Copenhagen, then took a train to the small Danish town of Hillerød, in the North Zealand region. It was a breezy, damp morning, and I was met at the little station in town by two fair-haired Danish beauties: Helene Hausner, a thirty-one-year-old doctoral student at the University of Copenhagen in the Department of Food Science, who greeted me with a dazzling smile, and Selma, her four-month-old daughter who was bundled up in a pram. I had come to learn about Helene's research into the complicated subject of how what a mother eats influences the flavor of her breast milk.

From the station it was a short walk to the center of town where we stopped outside a café facing a picturesque lake, in the center of which stood the Frederiksborg Castle, spread across three small islands. With slender spires, aqua-colored copper roofs and a Baroque garden, it was as fetching as any fairytale princess's castle. Built by King Christian IV, with construction lasting from 1600 to 1620, it was severely damaged by fire in 1859 and was eventually rescued by beer. The founder of the Carlsberg Breweries, J.C. Jacobsen, financed the castle's restoration on condition that it be turned into a museum, and in 1878 it became the Museum of National History.

Carlsberg Breweries was also the birthplace of one of the most familiar, universal science tools ever developed: the pH scale, the standard measurement of acidity, developed in 1909 by the head of Carlsberg Laboratory's Chemical Department, Dr. Søren Sørensen.

Settled in the café, Helene and I each ordered a milky coffee and Selma alternated between breastfeeding and sitting in her mother's lap, where, with the aid of soft toys, she became absorbed in honing her hand-eye coordination skills. Helene had spent four years studying sensory science and human nutrition, with a research focus on human eating behaviors. I had come to Denmark to visit her because of an article she had published in *Physiology & Behavior* in 2008, based on her study of how flavor molecules travel through the digestive systems of lactating mothers and into their breast milk. It was a flavor-to-milk journey that, up to that point, had not been studied much by scientists.

"I was surprised there's so little knowledge about how flavor compounds act in the body," she said.

In many cultures it is a common belief that some types of food eaten by breastfeeding mothers can affect their milk—that eating spicy food, for example, can give a baby an upset stomach. In Denmark, the two big forbidden foods for nursing mothers are chocolate and strawberries. Even doctors advise new mothers against eating them, as susceptible to old wives' tales as anyone. But after her flavor in breast milk research, Helene felt confident that nursing mothers could disregard this deeply rooted Danish belief.

In her study, Helene explored how caraway, licorice, banana and mint flavors were transferred from mothers' digestive systems into their milk. Her subjects were eighteen lactating women who ranged from twenty-three to thirty-five years old. They were healthy nonsmokers who didn't take medications or suffer from food allergies.

The data-gathering part of the study was completed over the course of three days during which milk samples were collected by the mothers two, four, six, and eight hours after they had swallowed four different flavor capsules after lunch. One capsule contained 100 mg of d-carvone (caraway flavor), one was l-menthol (mint flavor), one was 3-methylbutyl acetate (banana flavor), and one was trans-anethole (licorice flavor). The flavors were pure chemical compounds that had been prepared at the university's pharmacy and encapsulated in gelatin capsules. (I thought of Xavier and the possibility that if agar didn't work out to encapsulate drops of honey and mead, he could try gelatin.) Like Eves in the Garden of Eden, the mothers were allowed to eat anything they desired over the course of the three days, with the exception of caraway, mint, banana and licorice. One of the mothers, however, accidentally ate some cereal that contained dried banana, and she was soon eaten up with worry. But Helene (who thought all of the mothers in the study were terribly conscientious) simply didn't use this mother's milk on the day of the banana-in-cereal incident.

Although the study may sound straightforward, there was a lot going on at the molecular level. There were many factors involved, such as the way chemicals can either bond with or repel water, and the fat content of the mother's milk. Much of this was a mystery, even to Helene. According to her article, "the absorption, storage and excretion of flavor compounds in the human body are complex processes that are poorly understood." I personally found some of the flavor choices used in the experiment a bit bizarre—caraway and licorice, in my mind, are not flavors commonly liked by everyone—but they are widely enjoyed in Denmark. The flavors were chosen by Helene, not for their popularity, but because they represented different, basic molecular structures and are associated with a range of foods we eat—fruits, vegetables, candy and spices.

The results of her study showed that the licorice and the caraway flavors in the breast milk peaked approximately two hours after the capsules were eaten. Of these two, the licorice concentration was the highest—although it was actually only present in very low amounts. The mint flavor never peaked, but was present in fairly stable levels in the milk for up to eight hours. The big loser was the banana flavor, which couldn't be detected in any of the milk samples. But although the banana didn't show up in the milk, it did turn up some place else, and was the cause of the only side effect and complaint the mothers had about the study. They spent three days belching banana-flavored burps, all for the advancement of science.

According to Helene's article, there are several reasons why the banana flavor molecules may not have shown up in the breast milk. "First, the compound is rather volatile and was more difficult to dose accurately in the subjects. Second, the ester may have been hydrolyzed [broken down] in the acidic stomach environment . . . A third possibility is that the compound peaks before the two hour sampling."

Helene went back to the drawing board to determine whether the banana compound, indeed, peaked before two hours. And for this she asked an additional mother—a lactating friend of hers, not part of the original study—to ingest a capsule containing 100 mg of the banana flavor compound, and then collect milk samples one and two hours afterwards. In this case, Helene found small traces of the banana flavor in the milk samples. Her conclusion after this small, single-mother experiment was that fruit esters in general are transferred to mother's milk almost immediately after eating, but in extremely low amounts.

"There's hardly a chance a baby could detect the banana flavor," Helene said.

Helene's study confirmed the transfer of volatile flavors from a mother's diet to her breast milk; the flavors were, however, "transferred selectively and in relatively low amounts." To Helene, the big surprise was not that the flavors did, indeed, end up in the breast milk, or that the amounts differed depending on the flavor—but that there were so many differences in flavor quantity throughout the course of the three days in each individual woman's milk. According to Helene, this suggests that each woman's milk is "a continually varying flavor medium."

In Helene's study, she notes that another study showed that "infants whose mothers drank carrot juice during pregnancy or breastfeeding preferred carrot flavored cereals above water-based cereals." Yet another study seemed to show that bottle-fed babies are not as easily weaned as breastfed—implying that the bottle-fed ones were not as open to new flavors as the breast-fed, but that after they got used to the flavors, they were fine. As a scientist and experienced mother, Helene came to the conclusion that the food flavors that appear in breast milk in relatively low amounts possibly aid weaning—"Flavors in breast milk are a clever way for nature to help facilitate the introduction of weaning foods," she said—but are not the cause of any stomach upset in babies.

Although Helene originally met with some indifference when she applied for a grant for her study, it was eventually funded by the University of Copenhagen and her completed research created quite a buzz in the media. From a small article in *New Scientist*, a popular British science magazine, it was picked up by British newspapers, the BBC, and media as far as the Czech Republic and the United States. But although getting publicity was good, Helene was amazed by how many times her findings were manipulated and misquoted. One article in the British press reported that "nursing mothers have been shown to produce the human equivalent of a banana milkshake from their breasts." Another UK article included a photo that showed a close-up of a woman's large breasts, covered by a snug, low-cut outfit, but with no face to go with them. More appropriate for the Garden of Earthly Delights than as an illustration for a science article.

Helene never stopped talking to journalists, however, and was able to explain the study herself many times on Danish radio. When I met her, she had recently appeared on a Danish morning news program where she encouraged breast feeding mothers to eat as many different foods as possible.

Helene and I left the café to stroll through town toward the castle, where we were to have lunch. Along the way, she described her love of food as springing from sweet memories of the meals she had eaten at her grandmother's farm in Jutland, the northern tip of Denmark. Jutland is also the location of *Babette's Feast,* that gentle story (and film) of food and memory by Isak Dinesen. The pattern and caloric intake of a day of eating at Helene's grandmother's farm was based on a more active, outdoorsy life than many of us have now, and went something like this: Breakfast, mid-morning cake, lunch, mid-afternoon cake, dinner—finished off with cake. The love of food runs in Helene's family; she has three cousins who are chefs and an aunt who works in the dairy industry. But from an early age, Helene was more interested in science.

"I wasn't sitting around conducting experiments—but I was curious," she said.

That curiosity, the desire for puzzle-solving, had also manifested itself in her love of orienteering, a popular Scandinavian sport that combines cross-country running with navigation. She was serious about it for many years, and was a member of the Danish national team from 1998-2006. Orienteering is both a mental and physical endurance sport. Runners navigate with a compass, as well as a map provided by the organizers—there is no marked route to follow. The map shows hills, ground surface, and features such as boulders or cliffs, as well as the control points where the runners check in.

"Sometimes you don't want to take the trickiest route, even though it's shorter, because you might make mistakes if you're tired," she explained. "In the forest—everything is unknown," she said.

It seems perfectly natural that this love of finding her own way is linked to scientific inquiry. "People think scientists are in the lab all of the time," she said, but for her an enjoyable part of her study was being able to meet a group of nursing mothers and spend time with them, even when she had to drive two hours each way, in one case, to collect the milk samples.

Helene and I arrived at the restaurant, the Leonora, located in the former stables of the castle and named after King Christian IV's daughter. It was a serene place that combined old-world formality with crisp, modern simplicity. To begin our meal, we shared a plate of dainty herring from the tiny Danish island of Christiansø, with which we made our own little open-faced sandwiches of dark brown rye bread topped off with fish, minced onion, capers and crème fraîche. I guessed that the herring had been marinated with cloves which the waiter confirmed, but he couldn't tell us more, saying that the recipe was a secret known only to a couple of families on the miniscule island.

"I guess the clove will go into my milk, but I'm not sure if Selma will be able to taste it," said Helene.

We talked about Danish culture for a moment, and Helene tried to explain to me how important the concept of *hygge* (pronounced like *who-ga,* and translated loosely as "coziness") is to Danes. She searched for a good example, and looked around at another table where a group of diners were eating and talking intimately. Their table, like ours, was lit by a large ivory-colored candle. It was breezy and damp outside, but they—and we—were snug, chatting softly and feeding contentedly. A perfect specimen of *hygge*, which in Denmark is often enhanced by firelight or candlelight.

"Having a cup of tea with friends on a cold day—that's *hygge*," said Helene, trying to make sure I had grasped the important concept. It was, I thought, like those cozy tea and cake moments Proust and his invalid aunt shared on Sunday mornings, before Mass, when he used to go into her room and say good morning. Surely Proust would have been impressed that Danes have a word, and an obsession, for that sense of comfort, safety and pleasure that arises from the taste of good food eaten with good company.

After the starter we had our main courses. Asparagus salad for Helene and goulash soup for me, while Selma gummed some cucumber and asparagus.

"It's funny how she loves different tastes," Helene said proudly.

I discovered that one of the inspirations for Helene's study was the Danes' famous addiction to licorice candy. Like all true Danes, Helene's mother loved licorice candy and ate it while she was pregnant. Helene loved it and also ate it when she was pregnant, as well as now that she's breastfeeding, so Selma is no doubt getting a small taste of it, too.

Before beginning her flavor study Helene was doing some research on how difficult it is for adults to change their habits, not just in eating, but in general. One of her supervisors at the university pointed her in the direction of an interesting study: "Human Foetuses Learn Odours from their Pregnant Mother's Diet," by Benoist Schaal, *et al*, published in *Chemical Senses* in 2000. In this study, a group of newborns were "followed-up for behavioral markers of attraction and aversion when exposed to anise odor and a control odor." Those infants born to mothers who had consumed anise during pregnancy "evinced a stable preference for anise odour over this period, whereas those born to anise non-consuming mothers displayed aversion or neutral responses."

Helene sent me a copy of the study and I also watched a short video on the Internet about it that shows the scientist offering up handfuls of anise-flavored candy to the mothers in the study as if it were Halloween. Following that snippet is footage of a fetus slurping up its amniotic fluid over and over, and then seemingly wiping its mouth with the back of its forearm, like someone who has just taken a swig of booze straight from the bottle. While it may seem bizarre to imagine a fetus drinking amniotic fluid, the liquid is, in fact, humans' first drink, which they imbibe regularly in the womb. Helene said that when she was pregnant, it was during a scheduled scan that she, herself, saw Selma, a fetus, take the fluid into her mouth.

This study made a big impact on Helene. It offered proof that "during pregnancy there is learning taking place!" she told me.

In designing her own study, Helene had to make many decisions about what she would and would not ask her human subjects to do for her. In the study where breast-feeding mothers were fed carrot juice, the mothers had been asked to taste their own milk, and they, indeed, reported a carrot taste. But Helene thought this was asking too much.

"I've never tasted my own milk," she said. "And I don't think I'd want to."

However, she did do what many people have done in their own homes, when taking a cake or a batch of cookies out of the oven: she sniffed the breast milk samples. This natural curiosity gave her an unexpected clue as to the flavor of some of the samples. Although she couldn't smell any of the flavors in the milk samples, she could sometimes "feel" the cooling effect of the menthol flavor.

She asked her lab technicians to smell the milk samples and they, too, reported a cooling sensation when sniffing the menthol-laced milk. When I asked her how it was possible to sense the cooling sensation of mint without tasting it, she said I'd have to wait until the next day when I would meet a colleague of hers who could explain it all to me.

I left Helene and Selma at the station in town, and returned to Copenhagen. Part of the charm of listening to flavor experts in so many different fields was that it got me to thinking about lots of issues beyond what I had learned.

For example, although the flavors in Helene's study ended up in the mothers' breast milk in relatively small amounts, could it be, I wondered, that a baby's sense of taste is more sensitive than an adult's? And how would you measure that?

I remembered the first time I gave my firstborn daughter a taste of "real" food—pureed carrot—when she was around five months old. Up until that point she had only tasted breast milk—great stuff—but I remember vividly that each time I gave her a taste of the carrot from the little baby spoon she made a baby "mmmm" sound, and moved her legs as if she wanted to dance a little jig. This had been a private, single-baby experiment from which, after careful observation, I had drawn my own conclusions: Her experience of tasting the pureed carrot had been as thrilling as watching snow fall for the first time must be to someone who has never seen snow, or even heard of it before. I called home, and with my free hand held the telephone close to the action, and continued spooning, so that my mother could hear her first grandchild's first taste of real food, too.

TASTER'S CHOICE

The boy studied this manufactured substance, removed the wax paper, and took a big bite out of the chocolate-covered candy and again slowly chewed and swallowed. But again it was nothing—only candy—sweet, yes; otherwise, nothing, truly nothing. Once again the son returned to the father another substance of the world which had failed to bring him completion.

WILLIAM SAROYAN, *THE HUMAN COMEDY*, 1943

Early the next morning I met Helene's colleague, thirty-year-old Helene Christine Reinbach, at the University of Copenhagen. She was a petite, peppy scientist who had been studying the effects of hot spices on appetite, energy intake, and the sensory properties of a meal. I was curious to hear Helene Christine explain how, when sniffing, the sensation of cooling from mint flavor could be felt in the nasal cavity.

She sat me down at her computer for a Power Point demonstration, where I was introduced to the trigeminal nerve (fifth cranial nerve), which is defined in her PhD thesis as having "... free nerve endings going to the mouth, nose and eye region, [which] senses pain evoked by irritants, thermal, and tactile stimuli. Well-known irritants of the trigeminal nerve are capsaicin in chili, allicin in garlic, diallyl sulfide in onion, which can cause tearing and pain in the eyes and isothiocyanates in mustard, horseradish, and wasabi which cause tickling in the nose. Other common irritants are piperine from black pepper, cold, and warm temperatures, carbon dioxide from soda, zingerones from ginger, and cooling from menthol oil."

And so we are equipped with a nerve that not only picks up the cooling sensation of the volatile molecules in mint, but that also registers the burning, prickling, and tickling from a host of other foods as we chop, cook or eat them. Consider the possibility of having a triple-trigeminal experience in a single meal: Your eyes burn and tear up as you chop onions for chili con carne; your mouth burns while eating the finished, spicy dish; and you feel a prickling in your nasal cavity while trying to wash it all down with a carbonated soda. The trigeminal nerve has picked up all of these distinct and seemingly unrelated irritants on its three-way system, and transferred them to your brain. This nerve is so important, that scientists actually refer to it as the trigeminal sense.

"For us, it's completely a sense; but sometimes when you read about it, it's called 'mouth-feel,' so it's related to touch," Helene Christine noted.

She explained that our senses are divided into two groups: the so-called higher senses—sight and hearing—where it's not necessary to be in direct contact with what is being sensed; and the lower, or nearer senses (touch, smell and taste), where it is necessary to have chemical and tactile connections.

Given her research interests, I wasn't surprised to learn that Helene Christine loved chili, and was in the habit of growing chili plants in her living room. But during a recent apartment renovation, all of them had died. She told me about huge garden center where she bought her chili plants—extraordinary, considering that I thought chili only grew in sunny climates like New Mexico. These chilis—along with a range of other vegetables and fruits—were grown in greenhouses and were of various shapes, sizes, flavors, and colors with names such as Pinocchio's Nose, Monkey Face, and Satan's Kiss (small, red, and round). There's even one chili that tastes like lemon and black currant.

In the introduction to her thesis, Helene Christine writes that developing a taste for hot spices may be linked to the presence of other pleasant flavors that provide an environment which "transforms the initial painful sensations into pleasurable piquancy." I could see how this was as true with food and flavor as it is with art or music, or any number of human experiences. Rice and chili and chicken (or a single color or musical note), on their own, are limited: the rice and chicken too bland, the chili too spicy. But cook them up together (not forgetting garlic, onion, salt and a couple of tomatoes) and you get a symphony of blended, complex flavors along with tactile and visual contrasts—you get my mother's arroz con pollo—perhaps a clever way for nature to get us to eat a variety of foods, or even to encourage us to cook and eat together.

Before flying to Denmark, a food scientist I had interviewed for background research had sent me an article about a new restaurant in Brussels in which people eat a surprise dinner in complete darkness. As soon as I read about it I thought it might be a great place to take some scientists for dinner. For what exact purpose, I didn't yet know—but I thought it might be an excellent way to put into practice what I was learning about flavor and aroma. I happened to mention to Helene Christine that I was thinking about organizing such a dinner, but that I wasn't sure how playful scientists were when it came to exposing their sensory expertise to a public testing situation. I was surprised and pleased to learn, however, that some of Helene Christine's colleagues had already dined at similar restaurants in Germany. She suggested we join them at the department's weekly Friday pot-luck breakfast, just down the hall from her office.

We walked down the hall where about half a dozen of her colleagues, mostly women, were sitting at a table set up out in the open next to a window in the corridor, and they invited me to join them. On the table was a selection of breads, cheeses, jams, and honey from their favorite neighborhood shops, as well as from their recent travels. One of them asked me to try some cheese she had just brought back from Finland. I brought up the topic of dark-dining restaurants.

"I had a mild anxiety attack, and was thinking, 'how am I going to get out of here?'" a young scientist named Line Holler Mielby admitted about her dark-dining restaurant experience. "But I soon realized you can do things there that you wouldn't do anywhere else—like lick your plate."

Michael Bom Frøst, an associate professor in the Department of Food Science, Sensory Science, had also dined at the place. His main research interest is perception and cognition of food and beverages under real-life conditions, such as in restaurants, and he and Line had written about their dark-dining experience, which was included in the book, *The Kitchen as Laboratory: Reflections on the Science of Food and Cooking*.

Michael told me that his decision to work in food science had been inspired by three outstanding women involved in sensory science, and one of them was just down the hall preparing for a lecture: a Norwegian named Magni Martens, a senior research scientist at the Norwegian Institute of Food, Fisheries and Aquaculture Research, known as Nofima. Michael took me down the hall to meet Magni, a slender woman with short, gray hair. Although she was preparing for her lecture, she took a few minutes to sit down and talk with me about sensory science, which is a combination of chemistry, biology, and physics, she explained. And then she added, with Zen-like simplicity:

"Food is not nutritious before it's eaten." She then told me the story behind this intriguing statement.

In 1972, she was involved in a program established by the United Nations called the International Biological Program (IBP) at Makerere University in Kampala, Uganda. Many children in the area surrounding Kampala were suffering from *kwashiorkor*, a form of malnutrition that children living in poverty with starch-heavy, protein-poor diets may develop, often when they're weaned too early and replaced at the breast by a newborn sibling. One of the objectives of the IBP was to develop protein-rich food for malnourished children, and Magni was part of a group of researchers who had developed biscuits with the correct protein composition, which they brought into a nearby school for the children to try.

"We were really idealistic; we were optimistic that we could save the world," Magni said.

Nutritious as these biscuits were, however, they did not—could not—nourish the children for the simple reason that the children wouldn't eat them. They didn't like the taste.

The children were starving, "but the behavior, the faces of the children—it was clear that the biscuits were disgusting to them," said Magni.

"Food is not nutritious before it's eaten," she repeated.

The researchers had tasted the biscuits themselves, and had no complaints about their flavor. But they had not taken into account the cultural differences involved in taste and eating. It was after this startling, vivid experience in Uganda that Magni became interested in taste and the psychology behind why we like or dislike certain foods.

"And that was the start of my career as a sensory scientist," she concluded.

Magni and the other researchers never completed the biscuit study, since all foreigners were forced out of Uganda by then-president Idi Amin just three months after she had arrived. She went from Uganda to Kyoto, Japan for a year where she continued to study protein composition. Following that she received a degree in agriculture from the Norwegian University of Life Sciences, as well as a degree in philosophy and psychology from the University of Oslo.

Magni's story about the malnourished children of Kampala stayed with me for a long time. I found it almost impossible to believe that if I were starving I would refuse to eat what was on offer just because I didn't like the taste. There have always been stories of people in desperate situations who have forced themselves to eat the most unlikely and often revolting things in order to stay alive. So why hadn't the children eaten the biscuits?

Months later I found myself in the hospital recovering from emergency surgery to remove a perforated appendix. I had absolutely no appetite, and developed a strange aversion to both the hospital food, which was very plain but well-prepared, and food brought in by my family. Even simple white bread had a "smell." The nurses and my roommate tried to get me to eat more. My youngest daughter tried spoon-feeding me like a baby, but I sealed my lips and turned my head away. Days turned into weeks, and I knew that I must eat more—but if it tasted bad to me, I couldn't touch it. It was a nightmare, and I wanted to wake up and tell my family, "I had the strangest dream that I was in the hospital and you kept trying to feed me this weird meat that tasted like dog." It was only later that a doctor told me that the strong antibiotics I had been given were well-known for altering one's sense of taste. I had found out the hard way what Magni had discovered so many years before: incredibly, strangely, food needs to taste good before even a person who is wasting away will eat it.

SMELLY PIG

The flesh of no other animal depends so much upon feeding as that of pork. The greatest care ought to be observed in feeding it, at least twenty-one days previous to its being killed; it should fast for twenty-four hours before.

ALEXIS SOYER, *THE MODERN HOUSEWIFE OR MENAGERE*, 1851

The topic of unpalatable flavors continued in my next conversation. After meeting Magni, I chatted with Sandra Stolzenbach Nielsen, a twenty-six-year-old Danish research assistant I had also met at the breakfast. In contrast to all of the other scientists I had interviewed so far, her work involved, not researching or developing flavors, but masking them. She gave me copies of two of her articles published in the journal *Meat Science*. One focused on research to mask so-called "boar taint" that can be perceived in certain Swedish fermented pork sausages. The other was about how what a pig is fed can influence the perception of "boar taint" in meat from the pig. It turns out the problem is all wrapped up in a difference of the sexes: the sex of the pigs from which the sausages are made, and the sex and even the nationality of the people who eat the sausage.

Boar taint happens. It's a smell and a taste that can be detected in pork, both during cooking and eating, and it resembles the toilet area of a barnyard. But although boar taint can occur naturally in pigs, it can also be avoided. The traditional method is brutally simple: castrate the pigs destined to be turned into sausages and other pork products. But in recent years, the politics of animal welfare have gained traction. In 2009, Norway, for example, proposed a total ban on castrating male piglets. Alas, people still want their pork sausages—and here's where sensory science comes to the rescue.

The origin of the problem lies in the feeding, rearing, and sexual maturity of pigs, and some unpleasant chemical compounds, primarily skatole and androstenone. Skatole (3-methylindole) is produced in the large intestines of pigs. It smells like manure and accumulates in the fatty tissue. If present in high levels, it can be perceived when the meat is eaten. But here's where male and female differ. Although skatole is produced in both male and female pigs, only male pigs accumulate high skatole levels in their fatty tissue. Skatole is also rather mysterious as it doesn't appear to have any physiological functions. The second main chemical that contributes to boar taint is androstenone, a hormone that smells like urine. This one is produced in the testes of male pigs. It finds its way into the bloodstream and ends up in the animal's salivary glands.

The feeding study Sandra was involved with tried to solve the boar taint problem from the inside; that is, she and her colleagues were testing whether giving pigs a different diet would improve the taste of their meat. The pigs in the study were fed dried chicory roots and blue lupine seeds and the results were quite successful. The diet reduced the perception of boar taint in the meat.

Sandra's other study attacked boar taint from the outside—from the sausage point of view—in two ways: first, by fiddling with the starter cultures that instigate the fermentation process; and second, by smoking the sausages. In the first case, altering the starter cultures reduced the perceived boar taint but didn't eliminate it. Smoking the meat, however, made the sausage much more palatable.

But the research raised other questions: Why do some people—men in particular—not even notice boar taint? And why do some people actually like it?

Taste and smell are entirely private, inner-body experiences, but both consumers and scientists use language to help bridge the gap between what each individual tastes and smells, and what other people taste and smell. Androstenone was described by a panel during the study as having a host of unpleasant odors, including toilet, urine, sweat, and stable. Words used to describe skatole included stuffy room, cleaning product, burnt, and mold. But what happens if you can't smell any of these because of your gender? In one of Sandra's studies, the sensory panel was made up only of women, because the men who had been screened were not able to smell androstenone in low concentrations, a finding that was completely in agreement with other studies.

Personally, I don't like boar taint. I have sometimes smelled and tasted it in some pork products, and the best way I can think to describe it is "wet goat." Sometimes I buy a new type of sausage or cold cuts and don't discover the problem until I get home and taste. It happened one recent Christmas when I bought some salami from the butcher that I had never tried before. My eldest daughter and I tasted it, looked at each other—didn't say a word—and passed it along to my husband, in unison. From past experience, we knew he would eat it, no problem.

"Why are you giving it to me?" he asked. He always thinks we're playing a trick on him when we do that. When I explained, he took a small bite and said, "It tastes fine to me." Proof, indeed, that he couldn't possibly smell or taste what we could.

It seems that not only are there male/female differences in the perception of these smells, but there are also cultural preferences. Some people, it turns out, positively like a bit of stink with their sausages.

"In Denmark we don't like the flavor of boar taint, but people do in other countries—they've grown up with it," Sandra pointed out.

I thought of some well-known and well-liked cheeses, spices, and cooked vegetables that could also be described as having "bad" smells and tastes. Every time I open a jar of ground cumin before adding it to a dish, for example, I get a whiff of it and think "smelly feet"—but I wouldn't be without it. In just about everything, including flavor, humans enjoy contrast, novelty, and complexity, but balance as well as personal and cultural preferences also play a part—and with all of these factors to consider it's amazing to think that anyone ever gets up the gumption to create new food products, either for their local region, or for a wider, global audience. Or that Xavier was brave enough, back in Belgium, sniffing and tasting and trying to develop a special flavor for his mead that he hoped would be all things to all people.

APPLES OF YOUR EYE

We talked about the proper way to stew apples and how stewed apples are different from compote while our eyes rested upon the green apple that Picasso had painted for Gertrude Stein.

ALICE B. TOKLAS, *AROMAS AND FLAVORS OF PAST AND PRESENT*, 1958

After my trip to Denmark, I continued on my way, this time in search of someone who could explain to me how flavor gets into food we eat in its natural state. I decided to focus on the apple, one of the most popular fruits of all time.

It was an hour-long train journey from Brussels to Aarschot, a small town in Flanders, where I would meet up with a Flemish scientist and apple breeder, thirty-six-year-old Inge De Wit. When I arrived at Aarschot, I was collected at the train station by a rather intense, brooding taxi driver in a minivan, who began the short journey to Better3Fruit, where Inge worked, with a lightning-quick tour of the old town center.

We drove out of town, through pouring rain, to a quiet country lane leading to the offices of Better3Fruit, a private company of so-called "fruit designers" whose motto is "reinventing fruit for you." The Better3Fruit building was modern and functional, surrounded by hilly grounds covered in thousands of closely planted saplings. Seeing no reception area, I walked upstairs and wandered down a long corridor, cheerfully painted apple green, looking for Inge's office. I found her sitting at a computer. She had short, dark hair, and was dressed in casual trousers and a sweater, both the color of the wet earth outside. She made me a cup of coffee, after which we sat down in a conference room overlooking the orchard.

Better3Fruit, a spin-off company from the Division of Crop Biotechnics, Leuven University, began in 2000 to develop new fruit varieties for marketers, producers, and tree nursery specialists all over the world, starting with apples and pears and focusing on their disease resistance, productivity, and shelf life. But Inge, the company's breeder, cut straight to the heart of the matter: taste. First of all, consumers buy with their eyes, she said. An apple has to look appealing before someone will take it home and eat it. But what's the definition of appealing? In Europe, for example, northerners generally prefer green apples, while southerners mostly go for red. In the north, they generally like their apples to be on the sour, acidic-side, whereas in the south they often choose sweeter varieties. But, I wondered, do northerners buy green apples because they actually think they are more attractive than red, or because experience has told them that they are more sour than red? As Magni Martens noted in one of her papers, "The Senses Linking Mind and Matter": "What is the source of knowledge about an apple: Is it the physical measure of sugars, or the perceived sweetness, or both? Is it the detected sweetness on the tongue or the feeling of happiness when eating the apple?"

Inge explained that flavor in an apple is a combination of things: color, juiciness, texture, firmness, sweetness, and acidity—elements that can be measured and examined separately, but are expressed in different combinations in each apple variety. But how does flavor get into an apple in the first place? For that we had to begin with some botany basics. Inge used the Granny Smith, a popular tart, green variety, as an example. The Granny Smith came originally from Australia, and was named after Maria Smith who, with her husband, had emigrated from their home in England in 1838. According to an article written by John Spurway, a great-great grandson of Granny Smith, the earliest account of this variety appeared in 1924 in an article published in the *Farmer and Settler* written by Herbert Rumsey, a fruit grower and local historian in Australia. Rumsey had interviewed a fruit grower named Edwin Small who recalled that in 1868 he and his father had been invited by Maria Smith to "examine a seedling apple growing by the creek on her farm. She explained that the seedling had developed from the remains of some French crab apples grown in Tasmania. . . ."

Like the majority of apple trees, a Granny Smith cannot be pollinated by another like itself. If a bee happens to take pollen from a Granny Smith flower and deposit it onto the flower of another Granny Smith apple tree, fruit will rarely form, and thus Mother Nature avoids inbreeding.

"Sometimes, especially with hot temperatures during flowering time, some self-pollination can happen," explained Inge.

Most apple trees in the wild are pollinated by a variety of apple trees, all genetically different, resulting in apples that always look the same, but with children (the seeds inside the apples) that inherit genetic characteristics from both parents. In this way, apple trees are just like humans. Keep mating apple trees, or humans, and you will always get children that are different, not only from each of their parents, but from each other. In the case of apple trees, the process is intensified, since the trees almost always take multiple lovers. I thought of it this way: an apple is not the offspring of an apple tree. It is the sweet, tempting, edible mobile home of the seed-babies; a temporary embodiment of the mother that protects the seeds until they can break free from her—carried away by animals or humans—to be deposited elsewhere to flourish, independent of both parents.

Although I managed to get the seed/apple confusion clear in my head, I was still stumped. "How is it possible to have more than one Granny Smith tree if the original tree only bore seed-children different from itself?" I asked Inge.

The answer: you take a scion, a branch from the mother tree, and graft it on a rootstock. An arm of the mother tree grows on the body of another tree. From Granny Smith's branches come more Granny Smith apples. Grafting is a horticultural multiplication trick that has been practiced for thousands of years, not only on apple trees, but other fruit and vegetable plants.

Inge's work as an apple breeder is often about control. She allows Mother Nature to do her work, but she guides her, produces data on her, makes predictions about her. Inge and her colleagues are in the business of making what they consider better apples—better tasting, better at being stored, better at staying healthy—and for those reasons, one-night stands with nameless paramours cannot be tolerated. They match-make each pair, and pollinate the trees themselves, so they know who the parents of the seed-babies are. "To predict taste, aroma, and appearance, we need to know who both the mother and father are," she said. "When we cross two varieties together, we end up with around one thousand brothers and sisters, all with different genetic traits," she said. "Taste, aroma, color, acidity, and sugar content are determined by so many genes that it's difficult to predict what all these brothers and sisters will be like." They also prevent bees and the wind from getting there first by keeping the trees covered in clear plastic sheeting during the springtime.

Inge has a degree in agricultural engineering and a PhD in biotechnology. She has always preferred applied science, which is why she's happy to be planting seeds, grafting rootstocks, and of course, tasting thousands of apples. Sometimes she tastes a few hundred during the course of only a few hours. Typically from August to December Inge and her colleagues taste apples from their orchard daily. They also check the apples' flavors, textures, and aromas as they develop and change during storage. When Inge first began this job, she preferred sour apples like Granny Smith, but after having worked six years at the company, she has begun to prefer sweeter types.

"Perhaps it's because I'm getter older," she laughed.

After years of tasting countless apples she has learned that the sheer range of flavors they can develop is astonishing. Apples can taste like grapes, pineapples, lemons, and cherries—all quite nice, except for the pineapple taste, which Inge says is somehow all wrong in an apple. The pineapple flavor in combination with the apple texture, apparently, is what makes it unpalatable. And then there are other, more bizarre flavors.

"We've smelled and tasted aromas and flavors in apples that are so peculiar, there will never be a market for them; some even taste like cola—yuck," she said.

One variety they recently bred tastes like hazelnuts, which has divided people at the company; some love it, some hate it and so it hasn't come to market yet. Inge knows it's her duty not to focus on the apples she likes, personally, but to choose a variety of flavors for all different tastes. "I don't just choose my favorites, because I'm choosing new varieties for people all over the world," she says. Like Xavier, she was trying to choose flavors for people she had never met before, in countries she may never have visited.

And then there are the nasty off-flavors like vinegar, anise, and alcohol that often signal that an apple is overripe. Not surprisingly, this job can be hazardous to your taste buds. During harvest time when Inge tastes apples throughout the day, she can't eat other fruits because they taste like apples to her. In fact, the apple taste becomes so overwhelming sometimes that only chocolate or cookies can drown it out. Not that she minds.

Inge told me about one of their success stories, the Zari, which has bright red stripes on a yellow-green background with white lenticels (pores), and is the love child of the Elstar and Delcorf varieties of apples. Zari's story began in the springtime of 1988, before Better3Fruit spun off from Leuven University, when hundreds of Elstar trees were pollinated with Delcorf pollen. A few months later, the Elstar flowers developed into apples from which the seeds were removed and planted. The seeds became trees and, six years later, the first fruits were picked and tasted, and one was selected for its outstanding flavor. The company searched for a name for the new variety, and chose Zari, which means "golden rim" in Swahili (several of the company's apples have Swahili names). The first trees were sold in 2006. Zari's official description from Better3Fruit says that its shape is oblong-conical; it is sweet with a delicate aroma; it's very crispy and juicy; it has exceptional shelf life and storability; and it has good productivity.

Sadly, I was unable to taste the Zari. I have an apple allergy, a complex condition that can cause itching and swelling in the mouth and throat—the severity of which is influenced by time of year and apple variety. But like some other people who suffer from this allergy, I can safely eat apples as long as they're cooked, in compote or crumble, for example. But in thinking about how Inge said an apple must appeal to the eye first before the consumer will take it home and try it, I started thinking about how bored I've become with unnaturally perfect looking supermarket apples. In a push to cut down on food waste, however, vendors in some countries had been recently experimenting with selling misshapen, ugly, or imperfect apples and other produce that are just as tasty and nutritious. Recent scientific research has pinpointed why we enjoy slight variations in music, and so it must be with hunting and gathering. How much more stimulating cooking and eating would be if we were treated to more visual imperfections and oddities in our produce, as well as offered more heirloom varieties, and more chances to try new varieties, like Zari.

Inge and I left the conference room, and stopped off at a room down the corridor where she opened a family-size fridge that was filled with small clear plastic tubs—the kind you might buy to hold a salad for your packed lunch—all filled with wet sand, and around twenty-five thousand apple seeds. Next, we went outside to an enormous storage unit where the apples are kept in big blue crates labeled with tasting and storage dates. Entering the storage unit and seeing so many shapes and sizes and colors of the different apple varieties, all at once, was like stepping inside a giant Cézanne still life. These were not the monochrome red and green apples I had seen in my mind's eye when Inge and I had talked about the taste divide between northern and southern Europe. Here were apples with deep foundation colors overlaid with stripes and blushes; shades and tints of red, yellow and green, punctuated by off-white speckles, the freckle-like pores through which an apple "breathes." The Tunda ("fruit" in Swahili) apple is colored with a dark-red striped blush, except around the shoulder area where it looks like it has been dipped in lime-green paint. The shape and color of the Zonga ("embrace" in Swahili) is reminiscent of a ripe nectarine. Some of the apples don't have names yet. Apple 15 is described as having a beautiful smooth skin, bright red blush on a green-yellow background, sour-sweet, with "lots of aroma." The storage unit was like a giant sealed jar of tart, sweet apple scents. Apples were stacked up higher than my head, but I couldn't taste. I had entered the world of Tantalus, the Greek mythological figure who stole nectar and ambrosia from the gods and is said to have then revealed the secrets of the gods to his people. His punishment was simple: fruit and water, forever just out of reach.

BANANA-FLAVORED POP QUIZ

It is quite amazing the subtleties of flavour that our ancestors managed to create when you realize how few ingredients they had to play with.

ROBERT CARRIER, *GREAT DISHES OF THE WORLD*, 1963

From the flavor of nature, to the flavor of the genie in the bottle: flavor made in the lab, destined for food production. My next trip began, again, with a train journey—this time to a town called Hilversum, just south of Amsterdam. On the train, I met a man who was also headed to Hilversum. I told him that I was on my way to visit one of the manufacturing sites of a multinational called International Flavors and Fragrances (IFF). I apparently hit an olfactory nerve just by mentioning the place. It turns out he had grown up close by, in a house on *Slachthuisplein* (slaughterhouse square) that, during his childhood, had been plagued by a host of pervasive, unpleasant odors—not only from the slaughterhouse next door, where his father was the manager, but also from IFF.

"It was a sweet, penetrating, intrusive smell—not very nice," he said of the odors that had emanated from the factory.

As for the slaughterhouse—although the smells from there were deeply unattractive, he often hung out there because he was fascinated by animal anatomy.

"Maybe that's why I became a doctor," he said.

His name was Jan Stam and he was a professor of clinical neurology at the University of Amsterdam, and vice chairman of the department. We talked for a few minutes, during which time I learned that smell is the only sense that goes directly to the cortex. (The sensation of touch, in contrast, travels through the spinal cord to a relay station in the brain—the thalamus—where it's then sent to the cortex.) The olfactory nerve is also located close to the part of the brain that deals with emotion and memory. During this brief encounter, I had learned why certain smells can trigger a remembrance of things past.

I took a taxi to IFF and soon found myself at the locked outer perimeter of the company. My taxi driver buzzed and the gates opened onto a small road dotted with trees covered in white blossoms. On the short walk from the cab to the front entrance, I smelled nothing out of the ordinary. From the front door to the reception desk, I also smelled nothing. IFF, and the companies from which it sprang, had been in the flavor and fragrance business since 1833, and clearly the manufacturing processes, and their accompanying odors, had been contained since Jan was a child.

When I checked in at the reception desk, I was given a visitor's badge and a piece of paper outlining what to do in case of an emergency—something I would have expected at a nuclear facility, but not at a fragrance and flavor site. "In the event of an alarm: Leave the building directly after hearing the slow "whoop" sound and follow the instructions given by your host."

My host, Caroline Korsten, a senior flavorist in the sweet and beverages department, met me in the lobby where she discovered me sniffing musk at a hands-on display of decanted fragrances, sitting like works of art on small pedestals in the lobby. She was dressed neatly and trimly, without adornment. Caroline took me through to the flavor side of the building; flavor and fragrance, although under the same roof, were totally isolated from each other.

"They're two different worlds, kept separate," she said. "We don't want a contamination of the two."

We went into a conference room where she gave me a primer on the corporate world of flavor. Flavor is big business. IFF has offices all over the world, including India, China, Japan, the US, Europe, South America and Africa. In 2011 net sales were $2.8 billion. It began in 1851 with one William John Bush, a young Englishman who had studied botany, chemistry and herbalism. He became interested in the volatile components of herbs and spices, and at the age of twenty-two acquired a small factory in London where he began to distill herbs, spices, and woods to be used in tinctures and essences.

While the company may be global, the demands of its customers are often regional and cultural. When I visited, the company was doing some work in Russia where, Caroline said, even cream soda tastes different than it does in the US or Western Europe. "They have a cream soda, for example, that tastes more like caramel toffee." In just about every country in the world, she added, people like the taste of a different strawberry flavor. And in South Africa IFF has developed flavors from fruits as exotic as the baobab.

Flavor houses like IFF create flavors for a range of products including candy, dairy, and bakery products, as well as meats, soups, and beverages. According to the company's website, corporate customers "can choose from our library of hundreds of natural or nature-identical flavors—or we can develop an entirely new flavor profile to meet your exact specifications."

Caroline was well aware that the consumers' image of artificial flavors in recent years was not always a positive one.

"People think I work with unhealthy chemicals," she said, "but if I pick a real strawberry, maybe somebody has sprayed it with pesticide; maybe a dog has peed on it. At least I know my ingredients are 100 percent safe."

Being a flavorist, she said, is like being an artist. "When you're cooking, too, you play around until you have something that satisfies you," she said. "We balance experience and creativity with a customer's wishes. It's always a joy to create something that you later see on the shelf in a store or supermarket."

Caroline said she tests people's ability to identify a flavor—without using visual clues to help them—by giving them a candy that's been colored yellow, but that tastes like strawberry.

"Nine times out of ten, they don't know that the flavor is strawberry," she said.

Scientifically speaking, strawberry flavor has more than two hundred components. It's not practical, however, for a flavorist to use two hundred ingredients to produce any given flavor. The maximum number of ingredients in most flavors made at IFF is around sixty. Of course, the more experienced you are, the more adept you are at the principle of "less is more."

"I can make a very good banana flavor with only six raw materials," Caroline said. "But other flavors are much more complex." Coffee, like strawberry, is one of the more complicated flavors to get right. The trick, she said, is to be strict: "Do I need this raw material, or can I do without it?" she asks herself. "My philosophy is that I must create flavors with the minimum of raw materials."

It would be easy to say that manufactured flavors aren't always as good or accurate as flavors in nature, but the subject is far more complicated than that. Manufactured flavors can be just as strong and faithful as the originals, but perhaps they may lose something, or appear to taste different when they don't come in the same package as the original. How differently do we perceive banana flavor in ice cream or candy compared to the flavor of a real banana, the eating experience of which includes the pleasant feel and sound of ripping open the peel, and the sensation of the banana's unique, velvety texture. Flavors in nature aren't always ideal either. Take the tomato: Depending on the variety, and how they're grown, the flavors can range from insipid to glorious. The beefsteak tomatoes my father used to grow in our backyard vegetable garden were sweet, firm and full of flavor, not at all like the often disappointing ones we buy at the grocery store. The tomatoes of my childhood were blood-red fruit pods—more like a hybrid of meat and vegetable, fresh-picked from the vine. We often ate them in sandwiches on their own, just bread, mayonnaise, tomatoes, salt and pepper.

Next it was time to visit the sensory department, a large open lab. In one area there was a happy little group laughing, chatting, and passing out slices of fruit tart.

"They're celebrating a new win," said Caroline. A flavor had been created and bought by a client, which was cause for jubilation in this highly competitive business. A flavorist may create ten flavors for a corporate customer, and the customer may pick one—or none. And they only pay when they buy.

Caroline led me to a corner of the lab where she set up a little demonstration for me in order to illustrate how important it is that a manufactured flavor be in harmony with the other ingredients in any given product. She took two ordinary plastic cups and filled one with water, the other with milk. Into each cup she added a drop of liquid from a little glass bottle labeled "Magic Flavor." Then she asked me to sniff.

I took a whiff of the glass of water. It smelled like licorice. Then I smelled the milk, and it smelled green, like freshly mown grass. How could the same flavor from the same bottle smell differently in milk and in water? Caroline explained that Magic Flavor was indeed made up of two flavors: anethole (which gives licorice its flavor, and is the chemical that Helene Hausner used in her breast milk study) and cis-3-hexenol which gives off a green, grassy aroma, like freshly cut plant stems or leaves, and is a common ingredient in manufactured flavors. I was unable to smell the grassy aroma in the water because the anethole had overpowered it. But in the milk, the anethole had been absorbed by the milk's fatty acids, allowing the volatile molecules of the grassy aroma to emerge.

This demonstration is given to corporate customers as a vivid olfactory example of how important it is to know how a flavor will react with the other ingredients in a product. Fiddle with the ingredients in your product, and you may alter its flavor. "People forget the impact if they change the base," said Caroline.

Next Caroline handed me a lab coat, and ushered me into the Compounding Lab. To my surprise and delight, I was to be turned loose to make my own flavor.

Seven bottles of chemicals were lined up on the lab counter. There were also some pipettes, some thin paper strips like the ones used in perfume stores to smell different fragrances, a small, empty container in which to mix the chemicals, and a small electronic scale. The task was to make ten grams of banana flavor from the seven chemicals. Like an inexperienced golfer, I was being given a handicap. I hadn't forgotten that Caroline said she could make a banana flavor out of only six components.

Caroline left the room, and I was on my own. The first thing I decided to do was sniff and separate the chemicals into families. I dipped a thin paper strip into one of the bottles, sniffed, marked down the aroma that I perceived, and repeated this for each bottle. I ended up with three different banana aromas (sharp, sharper, and sweet/buttery); two spicy aromas, (nutmeg or clove, and cinnamon, I thought); a grassy/green smell (which, in this case, I thought resembled the aroma of banana peel); and vanilla (that was easy, as the bottle was marked, "Vanillin").

I tentatively put two grams of one chemical into my mixing container, then using the same pipette, I added one gram from another bottle. I realized, too late, that I had now contaminated one bottle of pure chemical with the contents of another. I could only hope that Caroline hadn't left me with some truly volatile chemicals. Earlier she had told me there was one chemical in the lab that, "if you drop one kilo of it, it would make the whole earth smell if it went into the gas phase."

There was a woman across from me, working steadily, quietly, and confidently. I felt as if I were at school again and it briefly crossed my mind to lean across the counter and ask another student for help. Instead, I started thinking about how much I had learned since I had started on this flavor quest. I recalled the banana beer I had tasted in Flanders and, as I did, I began to rise to the challenge. Looking at the bottles of chemicals laid out before me was like happening upon An De Ryck's beer, Jules de Bananes, broken up into puzzle pieces, and all I had to do was put it back together again.

A single flavor is made up of many components, except we don't usually think about that; we taste the finished product, not the individual cogs. In the same way, cakes and cookies are made up of various ingredients, but bear little resemblance to what went into them. (The basis of many cakes and cookies is, interestingly, nothing but rich food meant for growing young plants and animals: the whites and yolks meant to feed chick embryos; the endosperm [starchy tissue] of a grain of wheat [when milled, becomes white flour] intended to nourish the germ as it grows into a seeding; and butter made from milk a cow produces for its calf.) I sometimes make brownies by reducing the amount of sugar and butter, and replacing them with applesauce—pushing it as far as I can, without sacrificing taste or texture. This inspired my friend Shelley and her daughter Emma to make brownies in the same way, but instead of applesauce, they used pureed beetroot. The brownies were sweet, moist, and delicious, and I couldn't taste or guess the "mystery" ingredient. The beetroot had disappeared into a black hole of chocolate.

There was something thrilling about being back in a lab—and what a lab—after a lapse of decades. This was a power house where the magic flavors lived; a place where you could imagine holding a petite madeleine up to a prism, and a rainbow of eggs, flour, butter, sugar, and lemon zest would appear on the other side.

I started working faster, and with more confidence. In all, I mixed a three and a half gram combination of the three banana flavors with half a gram of vanilla plus one and a half grams of the grassy flavor and three grams of the spicy flavors. I sniffed. It was sickeningly sweet—banana intense. To correct this, I added half a gram each of grassy and the two spices and sniffed again. Of all of the flavors I had been given, I found the grassy one the most intriguing. It's a common chemical used in other fruit and vegetable flavor compounds, and great to tone down the sweet banana flavor, as I had just found out. Perhaps when we taste it in manufactured foods it subliminally reminds us that real fruits and vegetables spring up among the grassy-fragrant stems and leaves of the great outdoors.

Caroline returned. She made a solution of sugar water in which to taste my flavor. But as she was talking, she lost track of what she was mixing, threw it out—and started afresh.

"If you're not sure what you've done, do it again. You can't afford mistakes," she said.

She made up another batch and poured the sugar-water solution into two plastic cups—added a few drops of the banana flavor, and tasted.

I tasted, too. "Terrific!" I thought.

But Caroline said, "Too much spice."

She was too polite to gag, and in the name of accuracy, she tasted again.

"I can feel it burning my throat," she said.

She supposedly couldn't smell or taste much that day because she had a cold, yet my banana flavor looked like it was killing her. She was a highly trained, sensitive tasting instrument, and my banana flavor was as subtle as a sledgehammer. My mistake, she told me, was that I should have added a lot more green and not spice when I found the taste overly banana. She tried to cheer me up by adding, "If you bake an apple pie, it's going to taste different from my apple pie." My apple pie, I thought, would probably blow her head off.

Still, I tasted my Frankenstein banana flavor again, and again thought, "Not bad." Proof again that, no matter whether the flavor comes from nature or the test tube, taste is a subjective matter.

I received a brief botany lesson, quite by accident, when we visited Caroline's office near the sensory lab. Here she showed me a gift she had once received from a citrus supplier. It looked like a little box that you might see at a craft fair, small enough to fit into the palm of your hand. But instead of being an empty box someone might buy to hold a tiny memento, this was an empty box that, when I opened it, held a distinctive aroma that reminded me of the years my husband and I had lived in London. The place where we bought our first house, where our two girls were born, where we lost a baby boy who died in infancy. The box was brown, but not made of wood. It was made of the rind of a bergamot, a citrus fruit grown mainly in Italy. The oil from the peel is used to flavor Earl Grey tea, thus my association with London. The bergamot had been cut in half, all the flesh removed and turned inside out. It was then dried, molded and shaped. The outside of this strange container was lovely to touch, like a cross between ancient wood and leather.

I put the lid back on and sniffed the outside, the former white pith.

"That doesn't smell of anything," said Caroline.

Only the inside of the box held the aroma. Even though the whole thing had been thoroughly dried, the scent was still present because the bergamot oil lay undisturbed within tiny oil glands in the rind. These volatile molecules from the oil then escape into the closed chamber, held there until someone removes the top and releases them. It was as rewarding as lifting the lid of an old-fashioned hope chest, or cedar chest, as my mother used to call hers, and smelling the aromatic scent of cedar wood. I remember the last time we had rummaged through her cedar chest, she had pulled out the long, white baby dress I had been baptized in.

It was almost time to leave, but I had two more questions for Caroline. First, I told her about the dinner I planned to hold at the dark-dining restaurant in Brussels. I had decided that I might try a simple test during the dinner. Would the scientists be able to identify an aroma placed in the room secretly by me? I needed something like the bergamot aroma I had just sniffed: a ghostly scent, floating in the dark, disembodied from the food it was associated with. I needed a volatile chemical, the more volatile the better, easy to diffuse around the table. I asked Caroline what the most volatile chemical she worked with was.

"Acetaldehyde," she said, without missing a beat. "It has a low flash point [temperature at which it ignites]; it's dangerous; it goes straight to the brain and it makes you dizzy."

"We're not supposed to smell it," she added, in her deadpan manner.

Well, that wouldn't do for my experiment, I thought—both dangerous and not easily associated with a common food, like a tomato or a strawberry. But despite such a disturbing, frightening description, she later wrote to me that acetaldehyde occurs in nature in a lot of fruits, like grapes and star fruit—and also in apple juice, wine, and sherry. It's used in the flavor industry "to add juiciness and freshness" in mainly citrus flavors, but also in other fruit flavors. What I really needed was not a single chemical, but a chemical flavor like the banana flavor I had created.

Now for my last question. I asked if I could sniff something vile—specifically, indole, the fecal aroma that contributes to boar taint, and a chemical Caroline had told me she hates with a passion.

She looked it up on the database, located a sample of the chemical, and handed it to me to sniff on a paper strip. I hesitated—was I about to gag, or worse? I sniffed. It was bit unpleasant, but not at all what I had expected. To me it smelled more like mothballs, or stale cigar smoke. As I sniffed, it lost some of its power, and so I could stand to inhale more of it—and it sort of grew on me. I sniffed and sniffed, so I could memorize the scent. This appeared to horrify Caroline.

"You better not take that with you on the train," she said.

III. THE TASTE BUDS IN YOUR BRAIN

RECIPES FOR REMEMBERING

And he looked, and behold, there was at his head a cake baked on hot stones and a jar of water. And he ate and drank, and lay down again. And the angel of the Lord came again . . . and said, "Arise and eat, else the journey will be too great for you."

1 KINGS, 5–7

Back home in Brussels, I awoke early one morning. It was dark and cold out, and it took me a moment to remember that I was going to take a train to Wageningen University, in the Netherlands, where I had a lunch appointment at an unusual place called the Restaurant of the Future. It was a working restaurant where new food products and preparation methods, as well as consumer eating and drinking behaviors were studied, primarily for companies whose business was food related. My rendezvous was with a Dutch woman named Jos Mojet, a sixty-seven-year-old scientist who had been deeply involved in the startup of the restaurant. She had an MSc in social psychology and psychophysiology, a PhD in behavioral sciences and an interest in food and memory.

As soon as I boarded the train, I began to read a pile of Jos's scientific articles she had sent me that I had brought along for the long journey. Although "Taste Perception with Age: Pleasantness and its Relationships with Threshold Sensitivity and Surpa-Threshold Intensity of Five Taste Qualities" was, like most of the scientific reading I was doing, often way over my head, I nevertheless picked up one interesting tidbit in the "hypothetical explanations" section of the article: "People have a very precise memory for bitterness," she wrote. I started thinking about the word "bitterness," not only in terms of the sense of taste, as it was intended, but also in the emotional sense.

Indeed, I have a very precise memory for bitterness.

We were approaching Mechelen, a pretty Flemish medieval town with canals and a majestic cathedral, and I remembered that the surrounding farmland was famous for producing two bitter-tasting but much-prized vegetables: white asparagus, and *witloof* (white leaf), also known as Belgian endive. Although ghostly pale from being grown without exposure to sunlight, they are nevertheless visually appealing: the asparagus, ivory white, occasionally with a hint of purple blush; the endive, snow white, with celery-green or butter-yellow tips.

In one of my favorite recipe books, Ruth Van Waerebeek's, *Everybody Eats Well in Belgium Cookbook,* the author recalls her family tradition of going to her grandmother's house to eat the first white asparagus of the season. "We ate with our fingers (a rare and delightful treat), dipping the asparagus first in the melted butter and then in the creamy egg yolks." She also includes a recipe for Belgian endives, Flemish style, in which they're braised in butter and caramelized with a bit of sugar: "The resulting endives are meltingly tender with a hint of sweetness to balance the bitter taste."

As the train came into Mechelen I recalled a small museum near the station, the Jewish Museum of Deportation and Resistance, where I had once done some research in the archive department. The museum, next to the river Dijle, is in a corner of an old building with a large courtyard. Now condominiums, the building was originally a military barracks that was taken over by the Nazis in 1942 and turned it into a deportation center, the SS-Collection Camp Mechelen, for the whole of Belgium. In all, 24,916 Jews and 351 Gypsies were transported from this center to concentration camps in the east. It was during my visit to the archive department that an assistant archivist named Ilse Marquenie had given me an extraordinary cookbook: *In Memory's Kitchen: A Legacy from the Women of Terezin*—a collection of recipes written by the undernourished and starving women in the Czechoslovakian concentration camp of Terezin, also known as Theresienstadt. It's heartbreaking to realize that such a book had been written under such circumstances, and that it included many festive-sounding recipes, such as "Viennese Dumplings," "Rich Chocolate Cake," and "Rye Schnapps," all of which expressed, as treasured family recipes do, a past sense of happiness. One recipe called *Gefüllte Eier*, stuffed eggs with a variety of garnishes, was written by one of the primary authors of the book, Mina Pächter, and included a touching comment on how to prepare the dish: "Let fantasy run free." It was a unique book of recipes that had been published after a circuitous journey from mother to daughter. Just before she died at Theresienstadt, Mina had passed the "fragile, hand-sewn copy book" filled with recipes to a friend with instructions to get it to her daughter. It took a quarter of a century, during which time the book passed through many hands, before it finally reached Mina's daughter in New York.

The story of Mina sending such a book of recipes to her daughter got me thinking about how common, sacred and private these mother-daughter recipe exchanges—written or oral—must be all over the world, in all types of languages, cultures and circumstances. Perhaps there have been millions upon millions of such recipe hand-me-downs over the centuries, every one of them, I thought, similar to those Russian nesting dolls—if each doll held the same cake or plate of cookies, for instance.

But not all recipes get passed down. My nearly ninety-year-old mother-in-law recently told me she wishes she could find the recipe for a favorite chocolate cake her dear mother used to make for her when she was a child. She never thought to ask her mother for the recipe, and the only hint she had was that her mother used to cook from the Fannie Farmer cookbook, and so I offered to bake every chocolate cake in the book until we found it.

I continued on my way to Wageningen reading the rest of Jos's articles, including "Incidental and Intentional

Flavor Memory in Young and Older Subjects," where the subjects of Jos's experiment also remembered bitterness extremely well. Incidental and intentional flavor memory, although useful in a study about food, are drier, more restrained forms of their more flamboyant cousin, involuntary flavor memory, à la Proust: visceral memories sparked off spontaneously by the taste of something. Or bittersweet memories that flood back when copying down an old family recipe.

I remembered writing out my mother's recipe for dill pickles the year before she died, sitting around the kitchen table along with my daughters, who were also copying out recipes from her black, metal recipe box, cheerfully painted with yellow and pink flowers. And as I copied, I played a choppy, faded memory-film inside my head of our summertime "pickling days." First we picked cucumbers grown by my father in our urban backyard vegetable garden, a small rectangle of earth behind the clothesline. Cut to me and my sisters working under our mother's supervision at jobs along an assembly line in the kitchen: sterilizing the jars and lids in a shallow pan of boiling water; filling the jars with the cucumber spears and chips and sprigs of fresh, feathery dill, also grown in the garden; filling the jars up to the top with a boiled mixture, a ratio of three cups water to one cup of white vinegar and one-quarter cup non-iodized salt; and finally, sealing the jars, and labeling them with a date in what was, to us, the far future when they would be ready to eat. It was a lovely family moment, extracted carefully from an archeological jumble of childhood memories both pleasant and unpleasant, as most childhood memories are. Incidental memories, intentional memories, involuntary memories—there is an art to handling and storing them.

As the train pulled out of the Mechelen station, and we left behind the deportation museum and its dreadful past, and headed away from the Mechelen area farmland where *witloof* and white asparagus, with their distinctive bitter flavors, are grown, it occurred to me that surely most adults can remember at least one family recipe that, if enough time has passed, has become a sort of heirloom; a vehicle for revisiting and idealizing the past. Typed out on index cards, or handwritten on scraps of paper, these recipes are often forgotten for years until someone feels a sudden yearning for them. But unearthing such recipes is a ritual practiced not only by those who come from happy families. And so I had remembered copying down our family pickle recipe: sitting at the kitchen table with my elderly mother, keeping my children close—mentally on guard, as I had been throughout my childhood, wondering if my elderly father would interrupt, suddenly angry, or depressed, I knew not, as his moods could change in a heartbeat—and my mother and I reminisced. Six weeks after the pickles had been sealed in jars and stored in a dark cupboard, we tasted. The cucumbers were no longer bitter, but deliciously sour. The fresh colors had dulled, the crisp texture had softened, and damp threads of dill clung to them like delicate seaweed.

I reminded my mother of the time when, after one particularly long pickling session when we complained that six weeks was a long time to wait for a taste, my father, in a playful mood, suggested we go out to the garden with the salt shaker. We ran out to the vegetable patch, picked more cucumbers, rubbed off the tiny, black prickles from the waxy skin, rinsed away the dirt with water from the garden hose, and salted and ate them standing with bare feet on hot concrete under the clothesline.

PAVLOV'S RESTAURANT

The History of every major Galactic Civilization tends to pass through three distinct and recognizable phases. . . . the first phase is characterized by the question "How can we eat?" the second by the question "Why do we eat?" and the third by the question "Where shall we have lunch?"

DOUGLAS ADAMS, *THE RESTAURANT AT THE END OF THE UNIVERSE*, 1980

I arrived at Wageningen University where I met up with Jos at her office. From there we walked across campus to her regular lunch spot, the Restaurant of the Future, which had been fully functional for barely one year when I visited. It had cost over three million Euros (nearly $4 million at the time) and was apparently the first and only restaurant of its kind in the world. It is, in essence, a field laboratory for the study of the physiological, social and behavioral aspects of how we eat. It had the body of an ordinary restaurant, but the brain of a science institute. Everyone who eats at the cafeteria, including university personnel as well as visitors, is required to sign a consent form, which provides assurances such as "your conversation will not be recorded" and "the data and images will be used in research situations only, not for advertisements for the restaurant." On its website, the restaurant's objectives are "to explore and learn more about connections between behavior and physiology; liking and appetite; memory in food choice, perception and liking." And, my personal favorite: "To develop improved methods for measuring the effect of anticipation." Was this science or mind reading?

Jos and I entered the restaurant, which looked like a normal, medium-sized cafeteria, and was virtually empty at 11:30 in the morning. The sun was shining through the floor-to-ceiling windows and invisible waves of savory aromas filled the air. All eyes, in the form of ceiling cameras, were on us. Being, for the moment, one of only a few rats in the maze was not the least bit intimidating. I commented on the aromas. I looked around at the crisp, minimalist décor. I was relaxed and hungry.

We began at the coffee station, where Jos seemed delighted with my choice, strong coffee with steamed milk—*koffie verkeerd,* in Dutch—as if she had never heard of it, and ordered one also. Later, when we had lunch, I chose the same drink as she, a fruit-flavored milk. It's no surprise that dairy was a big part of the offerings at the restaurant—there were also a number of milk-based desserts—as we were in the Netherlands, where dairy is queen. But why, I wondered, did both of us end up drinking the same coffees and the same milk beverages? Did these choices reflect the fact that we liked each other, or that we were trying to please or impress each other? It was at this moment that I started down a slippery slope. I began to over-think everything.

When we had trouble working the coffee machine, a member of the restaurant staff appeared, and fiddled with it until it began to produce steamed milk.

"There!" she said in English, beaming at me. "The cow is coming!"

She seemed genuinely pleased to be helping us but I found myself slightly suspicious about her friendliness since I had read that the serving staff "can be ordered to be super-surly or obsequious as required." I asked Jos about this.

"In theory we could have them behave in certain ways," she said. "And I'm almost certain it would have an effect if we did."

We picked up our coffees, and she asked, "Where would you like to sit?"

I hesitated. What would my choice say about me? Jos had studied experimental and social psychology at Utrecht University. During her career, she had also been responsible for the creation of new research methods which make it possible to get a better insight into the unconscious motives and feelings and the role of memory in determining and directing food-related behavior. *Where would I like to sit?* I took a moment, then suggested we sit near the window that looked out onto the trees. But she replied that the tables there were too large for just the two of us, and steered me to a smaller table near another window.

"Would you like to sit in the booth, or on the chair?" she asked.

So many questions, so soothing a tone! I chose the more comfortable-looking padded booth over the sleek, modern chair but realized—too late—that choosing my comfort over hers surely carried with it some sort of Freudian, selfish connotations.

Jos and I soon returned to the buffet-style food counters for lunch where I managed, unconsciously, to choose lunch items that were mirror opposites of hers. Jos chose a richly colored, spicy oriental soup; I had a clear broth, with sauerkraut and sausage. She picked up an open-faced sandwich of dark brown bread topped with herring. I decided on a closed-faced sandwich of seeded light brown bread with smoked salmon. She took a small salad; I didn't have any. Neither of us had dessert.

As we got into line to pay at the register, where our food choices would also be recorded for scientific study, I noticed an ominous grey rectangle, flush with the wooden floor. It looked like a trap door in a horror film. But it was far worse than that. It was a scale used for weighing the customers at the register—not once, but every time they ate there. People sometimes have nightmares about being naked in public, but this was a new one, being publicly weighed—and judged, at least in my mind, over and over again. But just as we approached the scales, one of the employees noticed Jos standing in the long line, and took care of the check on the spot so we didn't have to go through the register. I was relieved not to have to step on the scale, but, as an example of how complex and contradictory the human mind is, I felt equally that I had just missed out on a significant group experience. I had been cut off from the herd.

When Jos and I returned to our table to eat our lunch, I hesitated. There were voices in my head saying:

"Sit up straight."

"Put your napkin in your lap."

"Wait for your hostess to begin eating before you start."

I tried to arrange my face so that it expressed calm politeness as I waited for Jos to begin eating. I thought about the short film I'd seen on the restaurant's website showing a so-called Face Reader—computer software used in the sensory consumer research labs above the restaurant—that could register six basic emotions on a human face. Based on Darwin's work, "The Expression of the Emotions in Man and Animals," the software focused especially on happiness and disgust. The emotional traffic on my face, however, must have reflected something more erratic—seizure-like, perhaps—because Jos looked at me with concern, and leaned forward slowly as if not to startle me.

"Would you like to pray?" she asked solemnly.

I laughed—not at the question, but at the realization that my expression was radically different than I had imagined. I picked up my spoon, and tried to save face, as it were, with one of the few words in Dutch I knew.

"*Smakelijk!*" I said, and Jos smiled. It's pronounced "smack-lick," and is the Dutch equivalent of bon appétit.

I took a spoonful of soup. *Delicious,* I thought, but I realized I had miscalculated and had ended up with a wet strand of sauerkraut hanging from my lips. I glanced up at the camera above my head, which could zoom in on anything or anyone it wanted.

The sandwich was no easier to navigate than the soup. The bread was harder than it looked, and I had dental issues which prevented me from chewing with ease on my left side. I seemed to lose all motor control. I talked with my mouth full, I repeatedly had to mop up bits of food from my lips and chin with my napkin, and crumbs rained down from my mouth onto my lap.

"I can never eat soup without having to use my napkin," said Jos, apparently trying, again, to mimic and express friendliness.

Jos and her colleagues are aware that what people *say* about their eating habits may be different from what they *actually do* when they eat. Which is why observing them in the restaurant—a consumer's natural habitat—is more useful than just making them fill out questionnaires.

Keeping human subjects in the dark during a study is also useful to researchers, as it produces more honest, unguarded responses. In Jos's study, "Incidental and Intentional Flavor Memory in Young and Older Subjects," the subjects were invited to take part in research on hunger feelings. Memory was never mentioned to the subjects during the study even though it was central to the research, and none of them recognized that memory had anything to do with it when they filled out a questionnaire after the experiment was completed. One of the topics Jos talked about during our lunch was the elderly, a group she has been interested in for many years, as reflected by the topics of some of her studies. The elderly population is growing, said Jos, and it's important to study them because "there's so much that can be done for them." The elderly choke easily when they eat, and they often produce less saliva, perhaps as the result of taking certain medications, which may interfere with their ability to taste.

Jos was born in 1942, in the Hague, during World War II when the Netherlands was occupied by the Germans. When I asked her how the war affected her and her family, she replied: "Compared to the thousands of people in the concentration camps, it was nothing, but my father died when I was six months old, and my mother had to feed me, and it was difficult." Perhaps I had been given a clue to Jos's interest in food and memory.

After lunch, we went behind the scenes. First stop, the video control room where the images from the ceiling cameras in the restaurant could be seen on a large screen that showed four areas of the restaurant at once. The people on the screen looked relaxed and natural. Unlike me, they seemed to have completely forgotten that everything they did was being recorded by the ceiling cameras.

Next, Jos gave me a tour of the sensory consumer research lab above the restaurant, where food and beverage companies come to assess their products, under all kinds of circumstances, for smell, color, and taste. There were many different rooms, grouped primarily around several main areas of study: the psychophysical labs, where swallowing, chewing, aroma, and taste are measured; the sensory lab, where products are tested by trained tasting panels; and the mood rooms where researchers studied "the effect of environmental factors on eating behavior and well-being (smell, light, sound, temperature, etc)"—and where the researchers could change, at will, the decorations, furniture, and lighting to alter the ambiance.

"It's a playground for me," said Jos, as we stood in one of the mood rooms. "Mood influences food, and vice versa."

During an experiment in the mood rooms, people might be asked to eat alone or together; by candlelight in a cozy setting, or with stark furniture and bare walls. The rooms are like empty stages, just waiting for props, scenery, and actors. One study Jos had headed on behalf of the Dutch Horticultural Marketing Board took place in four of the mood rooms, each with identical decoration. The goal was to study the influence that cut flowers have on people's moods while dining in a restaurant. One room didn't have flowers, but the other three contained various flower arrangements, only one of which had scented flowers. In two rooms the flowers were placed at the side of the room. In one, they were on the dining tables. The participants (sixty-two in all, eighteen of whom were male) didn't know the true purpose of the study. They were told they would be assessing human faces based on a number of characteristics, and that "the study was to take place around lunch time, as the facilities were available only at this time, and that this was the reason they would be offered a light lunch." The study confirmed what may have seemed obvious, that flowers in a restaurant are nice. But it was the link to memory that interested me. Jos's study concluded that "cut flowers have a subconscious positive effect on how restaurant customers perceive each other and have a positive influence on the memory of the eating event, thus increasing the chance of a repeat visit."

One particular finding from the flower study resonated with me, regarding my plans to secretly release an aroma during the dark dinner I was planning: "As the flower aroma becomes stronger, it will become increasingly difficult to concentrate and to relax, and with it an increasing level of aversion of cut flowers [develops]." I needed balance for my own restaurant experiment, I decided. The aroma I chose would have to be strong enough that there was a good chance everyone would be able to smell it, but not so strong that it would ruin the eating experience, and the memory of it.

TRANSPORT DELAYED

Yes, our actions are pictorial, our inventions are enormous, our thoughts are tragicomical, our temptations are burlesque, our desires are born of the flatlands, our paradises are made of dough and condensed milk, and our endearments are made of butter.

JAMES ENSOR (1860–1949), BELGIAN ARTIST

On my return journey, I changed trains at Rotterdam, and went searching for something to drink. I had plenty of time to have a coffee which I bought at an oasis right on the platform, a tiny café with room enough for a handful of tables and chairs. High up on a shelf next to the counter was a vintage leather suitcase displayed with a casually arranged bouquet of fake sunflowers, as if we were in the lounge of the Orient Express, headed for sun and adventure, instead of just commuting, or traveling on business. The managers of the kiosk had decorated their café with a bit of flair and fantasy—a real-life imitation of one of Jos's mood rooms, and an echo of the flower study—perhaps influencing customers to unconsciously think the food tasted better here than at other kiosks in the station.

In clinical-observation mode, I took a mental survey of what my fellow travelers (a group of Dutch students; a family speaking an Eastern European language I couldn't put my finger on; and two American businessmen) were eating and drinking. It was the usual stuff: soft drinks, coffee, potato chips, candy, sandwiches, and cookies. The food fell into three clear categories: savory or sweet; hot or cold; wet or dry. There were no subtle flavors here, no delicate aromas, but that didn't matter because we were cold and weary and needed a jolt from some strong, primary taste sensations. We wanted salt, sugar, and fat. We wanted food that would comfort.

Trains roared into the station, and roared out again, but none of them were mine. I had just heard an announcement that the train to Brussels was delayed. After finishing the coffee, I got back in line for something to eat and suddenly decided to buy a large package of chocolate-covered malted milk balls, which I hadn't had in years. I wanted to use them, not as a way to satisfy hunger, but as a circuitous, bespoke mode of transport—a means of taste-memory meandering. I had smelled and tasted malt during my two beer visits, which had reminded me that when I was a child I loved this candy. I especially liked the taste of the malt and its satisfying powdery crunch, in contrast with the vaguely rubbery texture of the chocolate. And after I finished college, left home, and started my first job in Washington, DC, my mother sometimes included them in the packages she used to send me.

I looked at the bag of candy in my hand and realized that perhaps my taste was slightly more sophisticated now and they wouldn't be as nice as I remembered. Besides, these weren't the ones I liked the best, although they were the kind my mother used to send me. The ones I really loved were the speckled, pastel-colored Easter eggs, as small as a robin's egg. On the outside, a hard, pretty candy shell, then a layer of chocolate, then the nugget of malt candy. These were the ones she used to buy along with other candies for the Easter egg hunts my father would organize for my sisters and me, and all the children in the neighborhood. We would line up, from youngest to oldest, and when he gave the signal, we would bang open the screen door and fly down the steps out into the back yard in search of candy. We would find a few chocolate eggs, often covered in colorful tin foil, sitting out in the open on the yellow metal seats of the swing set. Some were nestled in the knots and branches of the cherry tree. And others were surreally slotted into the narrow openings of the wire mesh pen he had built for our three ducks, one duck for each sister. He had bought us these ducks as chicks one Easter, and they had grown up into crabby, noisy pets that nevertheless supplied us with huge eggs, good for making cakes and cookies.

The anticipation, excitement, and joy of those Easter egg hunts far exceeded that of Christmas morning because the outcome relied entirely on your speed, your energy, and your ability to look in obscure places before anyone else did. It's what adults do, only they call it innovation; it was what Xavier was trying to do.

I stood on the cold train platform holding that ridiculous bag of candy, and thinking about the search for lost flavor, and how a fondness for certain foods can spring from the memory of small acts of kindness, from mother to daughter.

Had I filled out one of Jos's questionnaires at this point, I would have written that my mother had died the previous autumn, that she had had a long illness in which she had little or no appetite, increasing difficulty with tasting and swallowing, and had suffered chronic choking and severe dehydration. I remembered the visit home I had made a few weeks before she died. She had been mostly bed-ridden for months, but hadn't told me or my sisters, and it was a shock to find her looking as fragile as a baby bird, but with no idea, or desire to admit how ill she really was. Perhaps it's true, as Jos said, that a lot can be done for the elderly. But not if they won't accept the help they need. My mother did not want to go to a nursing home, nor did she want to hire anyone to look after her in her own house. At one point she actually said to me, "When I get better, we're going to travel."

Eventually I talked her into trying a Meals on Wheels type of service that would deliver supper for her and my father to the house every day. When the first ones arrived, I tasted the dinners to make sure they were good, and they were. I had ordered the special diet plate for my mother since some foods were forbidden to her in her condition. For my father, I got a heart-healthy plate. Although my mother could barely walk, we managed to get her out of bed and the two of them sat down at the kitchen table and started to eat. After a few bites, and without saying anything, they looked at each other—switched plates—and started eating again. I was horrified. What about my mother's special diet? But in the end, neither one of them ate much dinner anyway, nor did they like the dessert. I was sad, upset, and frustrated, but what was I supposed to do? Send them to bed without supper?

Every morning during my stay I would sit with my mother in her bedroom while she struggled to finish her preferred breakfast, a small bowl of Cheerios. One morning, as I was trying to get her to take a few more spoonfuls, my father came in from the backyard with a bouquet of zinnias from his garden which he had arranged in an old, cut-glass vase and placed on the desk where she could see them. He had always grown zinnias—a flower that thrives in hot, dry climates—and I have always loved their chubby flower heads, and their psychedelic colors: bubblegum-pink, popsicle orange and lemon-drop yellow.

"If you can't come to the garden, the garden will come to you," he said.

I wondered what Jos and her colleagues would have made of that flower-and-cereal scene, had they been able to observe it. How could they, or anyone, have accurately measured the impact, either on my mother's mood, or on my memory of the moment, that the humblest of flower arrangements had had on the simplest of meals?

Perhaps Proust could have written it up with the same sense of wonder in which he had described some particularly flavorful flowers from his childhood—not cut flowers in a vase, but dried linden blossoms used to make his invalid aunt's infusion. Proust was born in the late nineteenth century, so these were not the finely crushed linden blossoms packaged in tea bags that we know today, but wild-looking, loose tree material purchased from a chemist (pharmacy). In contrast to the restrained, careful words of Jos's early twenty-first-century study, "Cut Flowers Enhance Positive Feelings and Moods," Proust's passage is a sustained, lyrical description of the linden stems, leaves, buds, and blossoms that perfectly capture the bittersweet mood and memory of an otherwise ordinary eating experience. "Presently my aunt would dip a little madeleine in the boiling infusion, whose taste of dead leaves or faded blossom she so relished, and hand me a piece when it was sufficiently soft."

I changed my mind about eating the malted-milk candy and took it home to share with my two teenage daughters, who still enjoyed the little treats I brought whenever I returned from a trip. When I got home I emptied the big bag of candy into a pretty blue bowl. The sound of the malt balls dropping into the bowl was like the tinkling of wind chimes. I placed the bowl on the dining room table, turned out the lights, and called them in. I did this, I explained to them, because I was thinking about my next project, dinner in the dark. But they thought I was trying to trick them into eating chocolate-covered ants. What else would I have brought home from a place called the Restaurant of the Future?

We sat down at the table, in the dark, the only light coming through the windows from the Victorian-style streetlamp that stood in front of our house. We watched a man walking a dog, past the streetlamp, like an actor crossing a stage; his eyes straight ahead, never turning to look at the audience sitting in the dark. Around the dining room table, the colors had been drained from the tablecloth, the bowl, and from us, too. The girls were right. The candy did look eerily unappetizing. I lit a candle. "Smack-lick," I said, and we started eating the candy with more hush and sense of ritual than they had ever been intended for. The honeycombed center was just as flavorful and tactilely pleasurable as I remembered, but the chocolate, we all agreed, needed improvement. If only we could cover the nuggets of malt in Belgian chocolate. We began to list our favorite native brands of chocolate, including a certain boxed brand of individually wrapped rectangles of chocolate stamped with the image of an elephant raising its trunk. The girls prefer the milk chocolate ones, wrapped in red paper; I like the bittersweet dark chocolate ones, wrapped in black paper.

Perhaps it was the sugar rush, but I had a sudden entrepreneurial hallucination, plotting all the ways I could give my old favorite candy a complete flavor and appearance makeover. First, cover the malt in Belgian chocolate, some milk, some dark. Then, on top of that layer, a brittle candy shell in vintage blues, greens, pinks and yellows. I wanted a great combination of flavor, visual, tactile, and aural sensations. A lot of labor and thought goes into luxury Belgian chocolates. I had once wanted to get something festive to bring to a dinner party, and so I went to a fancy chocolate shop and bought a maxi-size, "limited edition" chocolate, as big as a cantaloupe—a replica of a molecule of chocolate, decorated with gold leaf. Anything was possible. So it was not hard to imagine applying some old-fashioned design details on the colored shells of my fantasy candy: tiny antique roses, or dinky polka dots, nostalgic design patterns all the rage at the moment on everything from tablecloths to teacups. These chocolates would appeal to the eye first, and when eaten, they would act like mini molecules of memory—gourmet mood candy for grownups.

In reality, the candy we were eating wasn't bad. We finished off the whole bowl, and as we ate I could hear the girls' crunching, plus my own crunching inside my skull. The sound and the feel of eating the candy was as pleasing and rhythmic as if we had somehow found ourselves strolling along a sidewalk laid with slabs of meringue in a film version of Candyland. We were safe and cozy, we had a little something to eat and share, and the conversation was light and sweet. A pocket-size helping of Danish *hygge*, lit by a single candle flame.

IV. SCENTS AND SENSIBILITY

THIS IS NOT A SOUP KITCHEN

Ancient Egyptian frescoes show us dinner guests with large cones of scented fat fixed to the tops of their heads; these were designed to melt during the feast, and drizzle deliciously down over the diners' faces and bodies.

MARGARET VISSER, *THE RITUALS OF DINNER: THE ORIGINS, EVOLUTION, ECCENTRICITIES, AND MEANING OF TABLE MANNERS*, 1991

Lots of things happen when we eat together. We smell and taste the food on our plates. We hear the rattle of cutlery and the sound of each others' voices. We recognize the faces and the expressions. The food may be warm, the drink may be cool. All of the senses are engaged. But what does it all add up to?

My husband and I often go on walks around our neighborhood after supper, and one of the joys of walking and talking is occasionally catching sight of other people in other houses seated around a dinner table. Warmly lit family portraits, framed by darkness. You cannot hear, taste, touch, or smell the *tableau vivant* behind the window. But you can catch a glimpse of someone's head slightly bowed over their plate as they cut their food, or someone else's animated face turned in conversation to the person next to them. The people around the table move naturally, and are unaware of us. The drama takes place at a distance, and is sometimes illuminated by soft lamplight or candlelight. No need for us to slacken our pace, or interrupt our conversation. In the seconds it takes to walk past the window, in a single glance, we have understood what it means to eat together.

Now, I set about turning this pretty image inside out. Turn out the lights, and turn up the volume on the sound, taste, touch, and aromas involved in eating together in the dark. My dinner at the dark restaurant would test my guests' sense of taste and smell. I invited Xavier and two scientists who had given me background information at the beginning of my research, both of whom worked for a beverage company: Sandra, who was petite and French, and Annelieke, who was tall and Dutch.

Sandra was a thirty-three-year-old sensory science manager who had had a variety of experiences during her career, including setting up and leading flavor training in Europe, Eurasia, and Africa, and co-authoring a paper on flavor generation in cheese curd by co-culturing it with selected yeast, mold, and bacteria. She also had an interest in theater improvisation and African dance.

Annelieke was a thirty-one-year-old product developer with a MSc in food technology. During her university studies she had had the exceptional opportunity during a traineeship in Iceland to develop a new, nutritionally enriched pasteurized "follow-up" milk for the relatively small population of Icelandic infants in this sparsely populated country. She was experienced and passionate about dairy products. "Each new product launch in the supermarket makes you proud as it is (partly) your recipe and energy. Between colleagues you can talk about 'your latest baby.' My mother sometimes goes up to strangers in the supermarket and says to them, 'This yogurt is really nice, my daughter made it.'"

Sandra and Annelieke asked if they could bring their husbands, and with me and Xavier this made six in all, a cozy number for dining in the dark.

Dark-dining restaurants are rare but becoming increasingly more popular throughout the world. A restaurant in Zurich called the Blindekuh (blind cow, which is what the Germans call the game Americans know as "Blind Man's Bluff") began in 1999 in a former Methodist chapel and claimed to be the world's first. It was opened by the Blind-Liecht foundation, whose purpose was "to foster the culture of blindness and promote dialogue and mutual understanding between sighted and visually impaired people." Some of these restaurants assert that eating without the sense of sight heightens our other senses. Whatever the reality, eating at a dark-dining restaurant had become a trendy experience fueled no doubt by its hint of mystery. It was the perfect place to play with your food, and for me to play a trick on a few friendly scientists.

I wanted to test my guests' sense of taste and smell as they ate their mystery meal in complete darkness. I would write a questionnaire for them to fill out after the dinner to see how good both their taste buds and their memories were. But within the straightforward questionnaire I decided to plant some questions that would prompt them to see if they could recall something in addition to the tastes and aromas of the dinner. Would they be able to smell, identify and remember a food aroma I would secretly place in the room? I followed the example of Jos, at the Restaurant of the Future, who always led her subjects to think she was exploring one thing, while she had actually set up the study to research something else. But I'm not a scientist so this couldn't be a real experiment—just a playful party trick.

I booked a table for six for the first Saturday in May at a dark-dining restaurant in Brussels, Only4Senses. I also made an appointment to meet one of the owners. Even though I wasn't attempting scientific method, I did want to check the place out so that I could prepare my little aroma test as carefully as possible.

The restaurant was located in the Galeries Royales Saint-Hubert, one of the oldest covered shopping arcades in Europe. Inaugurated by King Leopold I in 1847, it still housed luxury shops and restaurants in ornate surroundings under the original glass roof. It is also home to the first shop opened by the famous chocolatier, Neuhaus. According to the company history, Neuhaus began in 1857 as a "pharmaceutical confectioner" that had once sold "cough sweets, licorice for stomach complaints, and a bitter Belgian chocolate bar." It was also where the grandson of the founder spent months experimenting until he succeeded in creating "the world's first filled chocolate, which he named 'Praline.'"

Only4Senses was not to be found among the Galeries' other shops and restaurants, however, but beneath them in the renovated and expanded cellars of the Galeries, which now housed an eclectic little museum called Bruxelles en Scène (Brussels on Stage). Only4Senses is a here today, gone tomorrow sort of place that commandeers the museum for a week at a time to serve dinner. This bizarre eatery doesn't have a kitchen, or a cook, or its own food. Its meals are provided by a restaurant in the Galeries above it. I was reminded of the Belgian surrealist, René Magritte, he of the "*Ceci n'est pas une pipe*" painting, and thought a better name for the restaurant would have been, "*Ceci n'est pas un restaurant*" ("This is not a restaurant").

On the evening I went to check out the place, I strolled through the Galeries past an Italian glove shop, past a Belgian lace shop, until I found myself in front of the double doors leading to the underground cellars. The doors were wooden portals reminiscent of another era, and above them was a niche which held a plaster or marble bust of a man that, appropriately enough, was the spitting image of Dante. Patricia Raes, one of the owners, emerged to lead me down to the vaulted cellars where we started with a tour. The exhibits in the small museum were focused on Brussels' architecture, gastronomy, art, history, and culture.

Displayed against one of the cellar walls was a life-sized Brussels' street trolley, sliced in half lengthwise, like a lobster presented on a plate. There was also a photography exhibit loosely based on food-themes: a mother breastfeeding her baby in a park; churchgoers taking Holy Communion; and a group of people eating around a makeshift fire on an urban street.

After a quick wander around the museum, Patricia and I settled down in one of the rooms in which a bar had been set up on top of one of the exhibits, a sturdy piece of artwork that looked like a massive, elongated metal and glass pinball machine. As we spoke, the Only4Senses customers were dining in a pitch-black room at the back of this subterranean labyrinth, but I couldn't visit the room because Patricia didn't want to disturb them, or spoil my upcoming experience by giving too much away.

We chatted for a moment, when suddenly she was called over by one of the staff. She excused herself, and returned some minutes later, looking unhappy as she nervously sipped from a bottle of water. She told me that the whole restaurant, a pitch-dark room of around fifty seats, had been reserved for the employees of a single company, who had discovered that dining in the dark had released their inhibitions. Although I couldn't hear it, they had been shouting, chanting, and generally going berserk to the point that the blind wait staff was unable to function.

"I had to yell at them," Patricia said, "and tell them that

I'd turn on all the lights and kick them out if they weren't quiet. They were going crazy. I don't like shouting at people, but I had to do it."

That the guests were behaving badly was not really a surprise to Patricia. During her brief proprietorship of this new restaurant, she had observed enough to arrive at two empirically verifiable conclusions. First, when people eat in total darkness, they talk more loudly because they can't see the person they're talking to, and can't judge if they're being heard or not. And second, most people, especially when they all know each other, do things in the dark that they wouldn't normally do in the light—even though someone with an infrared camera is watching to see that everyone stays safe during the dinner.

"Basically the camera is there to make sure no one has fainted," said Patricia.

Or been murdered. As we were talking, Patricia's second appointment arrived: a Belgian woman who was doing some research for a thriller she was writing. The idea of eating in total darkness had certainly captured people's imagination. I headed back up the stairs out into the Galeries, and began walking toward the metro station, thinking about my visit. There had been something vaguely unsettling about Only4Senses. The corporate group that had behaved badly had given me pause. I wondered if this restaurant was too gimmicky. I walked into the the long pedestrian tunnel that led to the metro, preoccupied with thoughts about how I would design my aroma trick, and wishing I hadn't set it in a public place which was filled with too many uncontrollable variables.

Once inside the tunnel, and from some distance, I saw a crowd of about seventy-five people, mostly men, undoubtedly homeless, gathered in front of a couple of long tables where food and hot drinks were being given out by volunteers in red vests. The mood was festive and high-spirited. As I passed the long tables, I spotted a contemplative looking man, about forty years old, sitting alongside, yet apart from a group who had already been served their food and drink and were picnicking on a concrete ledge against the wall. The man was peeling a clementine. I thought it was sad that he only had one small clementine in comparison to the bowlfuls my family keeps in the living room at Christmas time when the clementine's flavor is at its peak. Part of the pleasure of eating them is to see which one of us can accurately toss their peels across the room into the fireplace.

"What's going on?" I asked him. I realized too late that this was a ridiculous question—I had never seen a mobile soup kitchen in the metro before, and what I really meant to ask was, "Is this a regular event?"

His face, thoughtful in repose, became pinched and angry. I thought of Darwin's book on facial expressions and emotions, which I had read after my visit to the Restaurant of the Future, and realized that my question had triggered a great surge of revulsion within him. He looked as if he had just eaten something disgusting.

He paused for a moment, sizing me up, and then answered in English. "We have nothing to eat," he said.

I didn't walk away or look alarmed. I asked him another question, and gradually we fell into conversation. It turned out he was an artist from Romania, and he hadn't been in Brussels long, but one of his bags, which contained his money and his passport, had already been stolen. He wasn't chronically homeless. He had once had his own house, and was a dab hand at barbecuing, he told me. But bad luck had plagued him, and suddenly he found himself at the mercy of handouts in a foreign country.

By now he had finished peeling his clementine, and offered me half.

"No thank you," I said.

"Take," he said. "Don't be proud."

I declined again, and so he began to eat methodically, enjoying each tiny section one at a time. Then he suddenly grabbed his bag from the ledge, unzipped it, and held it up for me to see. It was stuffed with clothing, atop which were a few candy bars.

"Take," he said, holding the bag under my nose. He was the host. I was the guest.

Again, I said, "No thank you."

We talked a bit more, and at one point one of the servers came around to collect the rubbish, and he helped her clear the nearby area. But soon the party was over, the tables and coffee urns were removed, the servers left for home, and the homeless simply left. I opened my handbag, took out my billfold, and gave him most of the cash I had on me. As an afterthought, I gave him some individually wrapped, slightly crushed *speculoos* cookies that I had in my bag. These are an old Belgian favorite, crispy spice cookies traditionally flavored with a mix of cinnamon, ginger, nutmeg, cloves, and white pepper, and often accompany coffees served in cafés and restaurants. I'm not naive enough to think that he was chatting with me without also hoping I would give him a handout. But I did believe from his facial expressions that he had genuinely enjoyed talking to someone, and that this talk was almost as important as the food he had been given. Our faces give away more than we think, as I learned at the Restaurant of the Future. And so we parted. And his eyes, which had burned as he described life in Romania before the fall of Ceaucescu, suddenly became as dull and void as a room in which the electricity has suddenly gone out.

I arrived home, went to bed without supper, and lay there thinking about the two groups of diners I had seen (and not seen) that evening. The people in the restaurant had been privileged enough to attend a private and rarefied dinner. Nothing wrong with that, except that they had behaved badly. Then there was the very public feeding of the homeless, which had had an extraordinarily convivial, party atmosphere in contrast to the reality of their situation. Europeans have had a rich and complicated history of food and cooking, as well as their share of hunger and starvation. Currently in Europe, and in other places around the world, there was a voracious appetite for all things food and cooking related: an avalanche of fabulous cookbooks and television cooking shows, as well as a vibrant interest in communal fruit and vegetable gardens, and pop-up restaurants—in contrast with increasing public concern with food-related health problems. I found it heartbreaking that citizens of countries such as Romania, recent newcomers to the expanded European Union family, could find themselves living literally hand to mouth. (And it would only get worse. By the time I finished writing, once middle-class people in cash-strapped countries like Spain and Greece would increasingly resort to such hardship measures as foraging for food in supermarket trash bins.) I had come to visit Only4Senses for a simple reconnoiter before my upcoming dinner, and had returned home thinking about how the subject of food touches on everything from entertainment, to politics, and to health and disease.

The dinner was coming up in the next couple of weeks, and I had yet to set up my aroma trick. I thought back to my visit to IFF, the flavor house in the Netherlands. I needed a volatile food aroma, and I decided that the best place to get one was from Sandra, one of the two scientists I had invited. So, with Sandra in on the trick, I now had one Dutch food scientist (Annelieke), one Dutch post doc in psychology (Robert, Annelieke's husband), one Belgian chef/mead maker (Xavier) and one French mathematics professor (Fabrice, Sandra's boyfriend)—and I wondered whether any of them would excel at detecting a rogue aroma in a dark room.

Sandra and I spent a lot of time discussing which liquid flavor would be best to use, and how to diffuse it in the dark. She put a short list of flavors together and did some tests on them. There were four in all: lemon, pineapple, cherry, and coffee. Eventually she chose the pineapple, and we decided it would work best if she deployed it during the dessert course. After a lot of discussion about how to release the scent without anyone being aware of it—it was harder than it sounded—I told her to simplify: put a tissue soaked in some drops of the liquid, and place it in the middle of the table. Don't wave it around, in case it touched someone, or caused a "breeze," and gave the game away. This was not ideal, but it was only a bit of fun, not a real scientific experiment. It was more like a Victorian parlor game. Even so, like many a scientist, I had a slight panic right before the dinner, wondering, "What the heck am I doing?"

DARKNESS EDIBLE

He started eating, I observed, in the normal sighted fashion, accurately spearing segments of tomato in his salad. Then, as he continued, his aim grew worse: his fork started to miss its targets, and to hover, uncertainly, in the air. Finally, unable to 'see', or make sense of, what was on his plate, he gave up the effort and started to use his hands, to eat as he used to, as a blind person eats.

OLIVER SACKS, *AN ANTHROPOLOGIST ON MARS*, 1995

On the night of the dinner, I was the first to arrive at the restaurant, and was surprised to see that the entry was partially blocked by a fantastically long table, probably long enough to seat 150 people, set up in the middle of the Galerie de la Reine. The tablecloth and napkins were black, and the plates, chairs, candles, and candelabras were golden. On either end of the table there were hundreds upon hundreds of red silk poppies planted on immense wooden display stands, just below nose-height. The Saturday night crowds wandered by, taking photos, and bending down to sniff the fake flowers. It made perfect sense, as I discovered this was a dinner given by a well-known fragrance company.

While my guests and I waited for the restaurant staff to let us in, other patrons also gathered, including a group that had a blind guest among them, a man who appeared to be relishing his role as unofficial group leader of the evening. When the doors opened we all descended into the underground area where we relaxed over a drink—in the light—before dinner. Everyone had news. Xavier's bees had arrived in Brussels, and he had set them up in Claire's mother's garden, but some of them had already broken away and formed another colony in a neighbor's yard. Sandra, whom I hadn't seen in months, was visibly pregnant. And Annelieke had just returned from working for a few months in France.

Our group was the first to be called to dinner. We were instructed to form a line, with me at the head. We were a human train, with both hands placed on the shoulders of the person in front. Our blind server, Anya, lead the way from the lighted underground museum through two sets of heavy, thick curtains, into a pitch black room. This was not like darkness when you're in bed, lights out, and you can still make out the shape of the furniture in the room. Here, there wasn't a pinprick of light coming from anywhere. It was more like a "Jonah in the belly of the whale" kind of blackout—an inky sea through which we were snaking our way toward our underground table.

We followed a narrow strip of carpet, laid on a stone

floor which I discovered when I veered off the edge at one point. We walked slowly in a somber procession. There was no noise at all, it seemed to take forever, and I had to fight the urge to ask, "Are we there yet?" Finally we came to a stop. "Reach in front of you," Anya said to me, "find the chair, and then slide down three more chairs and take a seat." I found myself at the end of a table in what felt like a little alcove, with a wall on my right and a low ceiling, both of which I explored with my hand and found they were made of rough, sharp bricks or stone. After much shuffling and calling out, I discovered that Xavier was on my left; across from Xavier was Sandra; and across from me was Fabrice. Unfortunately there was a mix-up and another couple was seated next to Xavier and Sandra, leaving Annelieke and Robert at the far end of the table, but this was soon sorted out. But it took a huge amount of effort during the dinner to connect with Annelieke and Robert, who were still one person away from me, one person too many in the dark.

Gradually the room filled up with the noise of other guests. There were perhaps fifty in all, and I began to notice dancing pinpricks of light here and there from people's luminescent watches. On the one hand, I wanted these people to remove their watches because the lights, tiny as they were, destroyed the total blackout experience. On the other hand, the friendly flickering was comforting. The air was heavy and a bit humid, and we could just as well have been in a forest on a moonless night with fireflies dancing around us.

Without being aware of what others were doing, but as if on cue, we each began exploring the area in front of us with creeping fingers, like musicians gently fingering a keyboard. We all discovered our first course, a salad, was already sitting in a round glass dish in front of us. We sniffed the salads. Annelieke and I both thought they smelled like peanuts. Beforehand she and I had been talking about what a nightmare a "surprise" dinner in the dark like this would be for someone with a food allergy. We dug through layers of different ingredients. There were no peanuts, but I guess we believed we had smelled them because we had planted the idea of peanuts beforehand. I asked everyone how they were tackling the salad. Some were using their forks in the usual way. Others picked the food off the tines with their hands. I was the only one who was fork-free, using my hands alone. To me, it was as natural and satisfying as picking popcorn out of a cardboard box in a dark movie theatre.

We all agreed that we were eating arugula with pine nuts, asparagus and tomato. Xavier later wrote on his questionnaire that he hadn't detected tomatoes at all because tomatoes at that time of year are not especially tasty. The asparagus was easy to identify because of its distinctive tips, hardly a surprise dish in the dark. Considering the food scientists and trained chef in my group, I would have preferred something more ambiguous. I tried to make the starter course more challenging by asking everyone whether they thought the asparagus was white or green, since both varieties were available at the moment. Everyone answered "green"—obvious, it seemed—because green stalks are generally thinner than white, and their texture and flavor are slightly different.

At one point during the salad course, Anya arrived with a bottle of mineral water, and the wine—one bottle of red and one of white—which Xavier and Sandra kept track of near their plates for the rest of the meal. Under Anya's instructions we served ourselves, in the way common for blind people: holding the glass with the tip of the index finger resting inside (we had tumblers, which were easier to use in the dark, instead of wine glasses). When the poured wine reaches your fingertip, you know the glass is full.

Soon our salad bowls were cleared, and the main course was served. This was a soup-like affair with pieces of meat—chicken, most of us wrote later on the questionnaire—and vegetables in a creamy coconut, spicy sauce, most likely curry. Only Annelieke later correctly wrote in the questionnaire that it was turkey, not chicken. Xavier, although incorrect, was quite sure of himself: "I discovered the chicken just by the structure and the noise that chicken makes between my teeth."

Again, I ate with my hands, picking out the meat and vegetables with my fingers, and then drinking the broth straight from the bowl. It was a wet, messy exercise that would never have appealed to me when eating with the sense of sight—but somehow, in the dark, it was not only far from disgusting, it was positively satisfying. I was giving in to pure, animalistic pleasure. But I was still anxious about how the dinner was going, and that mixed with lowered inhibitions made me an inconsistent host, and I hardly remember how I kept the conversation flowing. Again, there was nothing challenging, sensory-wise, about this poultry curry dish and I was beginning to feel quite disappointed in the dinner.

Suddenly, someone in the room screamed, and the room went deathly quiet. Time seemed to pause. After a tense moment, one of the wait staff assured us there was nothing to be alarmed about. One of the diners had just squealed for no particular reason. People started chattering again, and time resumed its normal pace. As we were finishing our main course, the waiters started singing happy birthday, and it was left to us to imagine that they were carrying a candle-less cake toward the birthday person's table. Or perhaps there was no cake at all. The entire experience we were having, the food we couldn't see, and the stop-start time frame in which we were eating it, was as jarring and unnatural as a loud dream with no pictures.

Although we laughed and chatted, the atmosphere around the table had an undercurrent of forced, unnatural intimacy. We had gone from a brief, get-acquainted drink—several of us were complete strangers to each other—to holding on to each other as we walked, touching hands as we passed wine and water bottles, getting fingers entangled over the bread basket. We were voices without faces, hands without arms. Conversational threads began, but went nowhere. We had expected to linger over the meal and the conversation, as Europeans generally do, but this dark dinner was a hallucinogenic variation of the lovely tradition. Time had lost all meaning. It seemed to drag one moment and speed up the next. I had difficulty completing sentences—I was sure I was boring everyone—without being bolstered by the usual conversational crutches of eye contact, smiles, and head-nodding. At one point Fabrice and Sandra started whispering and giggling and I momentarily felt as excluded, mortified, and enclosed without escape in our alcove as a teenager who's been ditched while on a double-date.

Soon we were asked to pass our bowls and cutlery down the table so Anya could collect them. Finally, it was time for something I was certain about: dessert and the release of the pineapple scent. Instead, Anya told us to stand up, form the human train again, and go out to the bar area where we would be served dessert. There were another fifty people coming in for dinner after our group, and everything was running behind schedule, so we had to move on. I was ridiculously disappointed that the little aroma test Sandra and I had planned wasn't going to happen after all. Our line formed, this time with me last, and meant to be holding onto Xavier's shoulders, but for a moment I couldn't find him in the dark, the "train" was leaving, and I was being left behind. Irrational panic set in until he found me, and placed my hands on his shoulders.

The journey back to the light seemed much shorter

and less dramatic than the way in. As I passed through the thick, double curtains back into the well-lit room, I was delighted to find Annelieke waiting for me. In the dark she had seemed so distant and cold, but here she was, her animated, intelligent self, full of lively conversation. And even more exciting was her first extraordinary remark. She wanted to know about the pineapple she had smelled. She was certain we hadn't eaten any pineapple, but had there been a tropical tree in the room, perhaps?

How bizarre, I thought. I was almost certain Sandra hadn't had time to put the pineapple-infused tissue on the table, but there was no time to find out how Annelieke could have smelled the pineapple, or if anyone else had smelled it. We had to eat dessert and fill in my questionnaire before the chef came out to reveal what had been in the "surprise dinner." We all gathered in the room with the giant "pinball machine," this time set with dessert, a bland-looking creamy affair served in clear dessert glasses.

We tasted. It was delicious, but the flavor was a mystery. This dessert—eaten with the lights on—turned out to be more confounding than the food we had eaten in the dark. Annelieke and Sandra correctly guessed we were eating *panna cotta,* cooked cream with gelatin, layered with ladies' fingers. Guesses as to its flavor, however, ranged from sweet vanilla (two people), apple and banana (two people), pear (me), and "don't know" (one person).

As soon as we finished the dessert, I handed each of them the questionnaire to fill out about the tastes and aromas of the dinner. When we finished the chef came out and listed all of the ingredients in the meal for us. Not much surprise there—we had all guessed most of it, except for the *panna cotta* which, it turned out, was made with mango and citronella. Annelieke's expression was priceless. She looked as if someone had just told her, on good authority, that the earth was flat.

But the biggest surprise of all came when I read the answers on the questionnaires and realized that not only Annelieke, but Robert and Xavier had also smelled pineapple. Under the question I had specifically planted for the aroma trick: "Describe any ambient smells or fragrances you perceived coming from the servers, the guests, and in general, in the room itself," I was amazed to read:

"Pineapple from Sandra!!" wrote Xavier. Robert had written that he had smelled pineapple in the room. "Pineapple—tropical tree ??" wrote Annelieke.

Under "write the name of each course and what accompanied it (even if you have to guess), and as many ingredients, flavors and tactile sensation from each you can remember," Annelieke had also written "pineapple in the air" under "main course."

I told everyone about the pineapple aroma test we had planned, and Sandra confirmed that, indeed, she hadn't had time to place the pineapple-scented tissue on the table before we had left the dark-dining area. "Then how was it that everyone smelled it?" I asked her, and she explained: the liquid pineapple flavor had gotten onto her hand when she had been fiddling with it before the dinner. All of that thinking and planning about how best to diffuse the aroma were of no importance. In the end, the problem had solved itself. The liquid had gotten onto Sandra's hands, which were not only warm, but constantly in motion throughout the meal, passing bottles of water and wine, reaching across the table for the bread, and using her hands to eat her dinner, each time leaving pineapple scent in her wake. In the end, she had become a human diffuser.

But why hadn't Fabrice or I smelled the pineapple? Perhaps he didn't notice anything unusual because he had been smelling pineapple at home or in the car on the way to the restaurant. As for me—perhaps I couldn't smell it because already planted in my mind was the certainty that it wouldn't be released until dessert was served.

I could see from their expressions that they were all exhausted. Although it wasn't excessively late, the entire experience had been so energy-sapping that the only reasonable thing to do was to return home as quickly as possible and sleep it off.

When we emerged from the restaurant, back into the Galerie, the perfume feast was in full swing. Seated at the fantastically long table were dozens upon dozens of mostly middle-aged and older women, dressed in radiant, jewel-colored dresses. Scores of golden candles glittered on golden candelabras. The conversations were loud and lively, the faces animated. After being in the dark for so long, this was a feast for the eyes, a Technicolor marvel.

At home I re-read the questionnaires, and my guests' overall impressions of the evening.

Sandra: "Very tiring in the sense that you have to pay attention with your ears. Ingredients were for me quite easily recognizable so not difficult."

Annelieke: "Wonderful. Cost a lot of energy focusing on voices/conversations. Easy to feel full/enough of food/cup was always big with no ending."

Fabrice: "Truly curious; one concentrates much more on taste and odor when one eats in the dark."

Robert: "Interesting/Intense."

And Xavier, always positive: "Wonderful experience."

As for me: I'm sure going to a dark restaurant would make for a really fun date with your significant other, or it would be a great night out with a few good friends. We know them so well, and it would probably be quite amusing. The group of mostly strangers I had invited had a lot in common, and had been chatting away with each other, in the light, before and after the dinner—but to feed and water them in the dark had been an unkind trick, in a way. I had cut them off from the facial traffic that would have given them so much information and pleasure, just as they were getting to know each other. The subtle, complex facial expressions we exchange with each other are the mealtime grace notes that help make dining together so satisfying. We come for the food, and stay for the company.

NECTAR & CO.

For nectar and ambrosia are only those fine flavors of every earthly fruit which our coarse palates fail to perceive—just as we occupy the heaven of the gods without knowing it.

HENRY DAVID THOREAU,
WILD FRUITS, 2000

Throughout my travels, I had kept in regular contact with Xavier. But after the dinner in May, I didn't hear from him again until August when he wrote to tell me that he and Claire had just gotten married and were on their way to South Africa for their honeymoon. The trip would include a visit to one of his international mead-making friends who lived there. This was appropriate, as mead, it is said, has long been associated with newlyweds.

Xavier sent me a note when he returned that read, "Concerning the mead, I hope the new batch is done, because the last one was a complete failure . . . I'm going to see Sonia on Monday."

I didn't understand. How could a beverage that had developed spontaneously, thousands of years ago, be a failure when it was being so carefully shepherded in a modern laboratory? But of course, that's the nature of scientific experimentation—to get the taste they wanted, new variables had been introduced, and there was more potential for things to go wrong.

We agreed to meet at Au Vatel one evening in early September, and despite his setbacks I was full of hope. I arrived at the café and took a seat at a table at the back, facing the door that led to the baking rooms. The café was about to close for the evening, and was uncharacteristically empty of customers. The noise from the street sounded muffled, and the bread shelves were bare. First Claire, and then a member of the staff offered me a drink, but I declined. I was convinced that everything had been sorted out with the mead, and I didn't want to drink anything before tasting it. After a few minutes, Xavier appeared in the doorway and walked towards me looking tan, relaxed and healthy. In one hand he held a red apple, which he took a bite of, and in the other hand he carried a small glass jar. Inside was a translucent, golden liquid. We exchanged the usual conventional Belgian kiss on the cheek in greeting, and I wouldn't have been at all surprised if, at that point, he had said, "Behold, I bring you tidings of great joy."

Instead, he said, "This is for you."

I had started this journey with him just under a year earlier, and here was the moment I had been waiting for. He sat down, opened the jar, and said, "taste," and started to tell me what it tasted like. Then he stopped himself and said, "You tell me what it tastes like."

It was like one of those wrenching dreams you have

when you discover something strange and rare which, in the dream, is the sole object of your desire. But as you reach out for it, it suddenly disappears, and you cry out in your sleep. The type of nightmare Tantalus probably suffered from on a regular basis. I looked more closely at the jar and realized that Xavier was offering me, not his mead, but a sample of honey from his bees. From afar it had the color of some of the meads I had tasted at the Slow Food dinner. I had been so positive I was going to taste the mead that my eyes, and my brain, saw what they wanted to see. I was reminded that one of the goals of the Restaurant of the Future was "to develop improved methods for measuring the effect of anticipation." Nearly a year of excessive anticipation had corrupted my ability to see what was right in front of me.

Nevertheless, I dipped my finger in the golden liquid and tasted. It tasted buttery in the beginning, and after a moment, also minty. It was a sweet, delicately perfumed liquid that went down with a smooth, cool sensation. It was like honey on ice, and it had a livelier trigeminal zing than any store-bought honey I had ever tasted.

"You'll never guess what kind of honey this is," said Xavier, dipping a little teaspoon into the jar and tasting it himself. I thought for a moment, but before I could answer he said, "It's linden honey." Coincidentally, the nectar from which the honey had been made had come from the same type of tree blossoms from which Proust's aunt's tea had been made.

"But what about the mead?" I asked gently.

Still not ready, he told me. No matter how pleasurable it was to taste the honey, I was deeply disappointed that I wasn't going to taste the mead that evening. And I couldn't help wondering whether I was ever going to taste it.

Xavier stood up and went behind the bakery counter and brought out more honeys for me to taste, including a handsome little jar labeled "Miel de l'Etna," an aromatic, creamy orange blossom honey from a village near Mount Etna in Sicily. It smelled like the orange blossoms it came from and tasted sweet at first, but tangy on the way down. And it had the pleasing color and texture of sugar and eggs when they've been creamed together to make cakes or cookies. This was one of the honeys that had been tested in Sonia's lab as a possible ingredient for Xavier's mead, and Xavier had also begun importing it and selling it to gourmet shops in Brussels.

"I want to show people the huge range of aromas available in different honeys," he said. He was building a honey empire, the crowning glory of which would be his mead with its elusive flavor.

In all, fourteen different types of honey, and combinations thereof, had been tested in Sonia's lab for possible use in Xavier's mead. They ranged from common types like acacia, to more unusual ones like hawthorn, bramble, black alder and strawberry tree. But most intriguing of all was honey made from a sticky, sugary liquid called honeydew, which doesn't come from flowers at all, but is secreted from a winged insect called the Flatid Planthopper (*Metcalfa pruinosa*). Honeydew is also produced by aphids and other small insects that, like the Planthopper, feed on plant sap. This aptly named liquid food is collected by bees, butterflies, and ants who "milk" the aphids. Even geckos like it, and eat it in the most extraordinary way: they have been known to stand next to a Planthopper in order to catch a little ball of honeydew in their mouths as it's flicked from the Planthopper's derrière. This symbiotic relationship, known as trophobiosis, works as a barter-style exchange in which food is offered as payback for protection. Bees don't appear to have this relationship with sap-sucking insects, but are said to gather honeydew from wherever they can find it when nectar is scarce.

Xavier was still full of enthusiasm for the mead project, and assured me that he would soon have a drinkable batch, perhaps in three or four months.

"Maybe it won't be my dream mead, but it'll be a good one," he said.

The plan was to put something on the market that was not too expensive, as he continued to fine-tune the flavors and aromas behind the scenes.

"The world of mead is moving forward," he said, "and other people are researching it around the world. If I wait too many years someone else will find what I'm looking for." In Xavier's mind, this was turning into a race to the moon, and he was determined to be way out in front of those who were making a significant contribution to the science of this obscure beverage.

"The scientists," he continued, "can explain the scientific processes. They have had experience with beer and wine, but no experience with mead. They're the scientists, but I'm leading the research with my taste buds. I know many things about mead and I can tell them when a taste is wrong."

We talked until long after the bakery had closed and then got up to go around the corner to Au Vatel's take-out shop, which was open throughout the night. It was a beautiful, warm, autumnal evening. We walked onto the square where people were eating and drinking at sidewalk tables, and where the queue at the Friterie Antoine was growing longer by the minute. At the corner we turned left onto Rue General Leman, and re-entered Au Vatel through a dark, narrow corridor until we emerged in the baking rooms where I had once heard the baguettes singing. I followed Xavier past the customers lined up at the sales counter, and into the area where the racks of freshly baked bread stood. He selected a good-sized, round loaf, rapped it on its underside and, satisfied, put it in a bag for me as though it were the pick of the litter. It was a rustic white loaf, the golden crust of which was covered in attractive dimples and a light dusting of flour. I asked him to choose a large baguette for me, also.

Back on the street, we parted, and he promised to keep

in touch. On the way to the metro I carried the baguette under my arm, and the round loaf in a brown paper bag marked "Au Vatel, anno 1950" with a picture of Claire's grandfather pedalling a bicycle-like contraption with a huge container in front for transporting bread. The baguette was enormous—around three feet long, and fat enough that my fingers didn't touch when I held it around its middle. Out of habit, I pulled off the hard, knobby bit at the end, ate it, and then continued to pull off bits and eat them until I reached the metro. In my family we usually buy two baguettes when we walk down to our local bakery for bread—one to take home, and one to eat, as we walk and talk. It starts out with removing the pointy bit at the end, just for a taste, but nearly always ends with most of the baguette being eaten. It's impossible to resist if the bread is still warm—and we have decided not to feel guilty about it anymore.

The one Xavier had picked out for me was the granddaddy of all baguettes, and it was my supper. I had eaten nothing that night but bread and honey—and had had nothing to drink. I couldn't stop thinking about the mead, and what it might taste like.

"I'm searching for something that may turn out to be impossible," Xavier had said to me before I left. "For me, mead is nectar. Nectar from flowers becomes honey, and then is turned back into nectar—mead—again."

A few weeks later, I again received news from Xavier.

"The last batch was contaminated by bacteria. But we

found the clue and the two next ones will be ready in three weeks."

But who was to say what could go wrong with that batch? Where would it end?

The autumn passed, and I heard little from Xavier. The first anniversary of my mother's death had come and gone and, coincidently, my husband and I had spent it at the home of some friends, Chris and Karen, to celebrate the life of Chris's mother, who had recently died in England. She was an Armenian who had met her English husband while living in Jerusalem, where she had also survived the bombing of the King David Hotel in 1946. When I thought about it, she was the start of the six degrees of separation that had led to my accidental meeting with Xavier. She had survived the bombing, married, and had Chris. Chris and his wife Karen had introduced me to Jacky, their neighbor. Jacky had introduced me to her friend, Hughes, and Hughes had introduced me to Xavier.

Chris's mother, Vicky, had been petite, feisty, and a terrific cook. She had let me cook with her twice when she had been visiting. I wanted to bring something to the afternoon tea Chris and Karen were hosting to celebrate her life, and tried to recreate her recipe (she cooked like my grandmother; her recipes were all in her head) for *maamouls,* a Middle Eastern disc-shaped confection the size and shape of a hockey puck. The dough is made with semolina, butter, yeast, and sugar; the stuffing is made with crumbled walnuts, cinnamon and sugar. I borrowed one of Vicky's special *maamoul* molds from Karen and made them as best I could (with a little help from the Internet), and brought them over early, while they were still warm. I knew they were not the same as Vicky's, but Chris tasted, and was extremely kind about them anyway. Like petites madeleines, a mold, traditionally made of wood, gives *maamouls* a decorative geometric design in relief, as well as fluting around the edge. And like petites madeleines, they inhabit that ambiguous, tasty netherworld between cake and cookie.

By November, Xavier had written to me again saying that there was nothing new concerning the mead, but that he and Claire were worried about her grandfather, Jean-Baptiste Wyns, Au Vatel's founder, who was gravely ill. Meanwhile, my eighty-one-year-old father, unable to cope after my mother had died, had entered a nursing home where one of his main health complaints was chronic dehydration. Crabby by nature, he was stubbornly refusing to drink enough—but inconsolable that his mouth was constantly dry, and that food no longer tasted good.

Christmas came and went. Then at the beginning of January I received a chirpy email from Xavier wishing me a happy New Year, and letting me know that he would call me after his meeting with Sonia the following week. Then nothing for a few weeks. So I contacted him, and he said we could meet at the beginning of February. But he didn't mention the mead, and I tried not to anticipate another disappointment.

He was always extremely busy with work, family and his mead project, so we agreed to meet for an early breakfast. On the day, I left home at 6:30 in the morning and traveled in the dark by tram and metro across town to Au Vatel. For weeks, Belgium and the rest of Europe had been suffering on and off from freakishly harsh snow storms, but at the moment the streets were clear and by the time I had reached Place Jourdan, day was breaking. Au Vatel had just opened, smelling of butter, yeast, and coffee and was cheerfully decorated for Valentine's Day. The shop was alive with the morning breakfast crowd, and I sat in my usual spot in the back at a table facing the door to the baking rooms. As I waited for Xavier to appear, I watched one of the employees, a dark-haired woman, carrying freshly baked fruit tarts and other pastries on a large tray to the display cases near the front of the bakery. Back and forth past my table she went, leaving a rich scent of butter and sweet, cooked fruit in her wake. Her quiet concentration, and the way she held the heavy tray of tarts reminded me of the food bearers in Breughel's famous painting of a peasant wedding feast.

Xavier soon walked through the door, greeted me briefly, placed a square-ish cardboard box on the table, and went to get us some breakfast. The box was completely ordinary—just large enough to hold a toaster, or some similar quotidian item. I leaned across the table to peek, but the flaps on top were tucked into each other so tightly I couldn't see inside. He soon returned with breakfast: *lait russe* for me, freshly squeezed orange juice for him, a bottle of mineral water and a plate of small bread rolls, croissants and *pains au chocolat* to share. As usual, he was dressed casually, and on this particular morning he wore a thick winter scarf wrapped around his neck. Although in his late twenties, and over six feet tall, he always looked eternally clean-cut and boyish to me. I could imagine him as a Boy Scout, as he was when he was younger. He drank most of his orange juice in one long gulp, and then told me that Sonia's job was finished, which immediately raised my hopes. *What was in the cardboard box?*

He continued talking, never acknowledging the mystery parcel. The problems with bacteria, which had somehow gotten into recent batches, had been solved, but the mead was still watery, and now he wanted to turn to experts in the wine business to help him give it more body. He praised Sonia and her team for getting him this far, and for achieving the main goal of the project: retaining the aromas from the honey in the finished product.

"Sonia had the best machines of anyone for analyzing aromas," he said, "including a huge machine with a 'nose' on one end and a printer on the other."

I drank my coffee, but I was too excited about what might come out of that box to eat anything.

"I learned a lot," he said. "Before Sonia, everything was hypothetical, all on paper, but she showed me everything."

Xavier had not only received expert guidance for his project but, in the tradition of a medieval apprentice, had gained valuable first-hand scientific experience. He tore a *pain au chocolat* in half, ate it slowly, and turned the conversation to the future. Mead is a drink best paired with food, as well as to be used as an ingredient in cooking, he said. His next move was to test it with a panel of chefs whom he hoped would develop some interesting recipes for it. I admired his forward thinking. He realized that his mead should not stand alone on a pedestal. I recalled the wonderful lamb marinated in mead that he had made when we first met at the Slow Food gathering.

Xavier finished his *pain au chocolat*, brushed his hands together to remove the crumbs, and—my heart gave a little leap—opened the cardboard box and reached inside.

"And now we taste some honey," he said as he placed a variety of jars in front of me. This was odd, as well as disappointing, because that's what we did last time.

And so we tasted. Honey from his bees and from Zafferana Etna again, and from the Belgian border with Germany, made by his first beekeeping teacher. There was dark honey, golden honey, solid honey, liquid honey, multifloral honey, linden honey, white lavender honey, chestnut and pine tree honey. Tasting honey with Xavier was an act of slowing time that was as pleasant as when I remember spending a quiet, concentrated moment with my children when they were very young, admiring and sorting through their baskets of richly colored, dyed Easter eggs, until the memory of plain old white chicken eggs—or in this case, ordinary honey—was erased from my mind. People were eating breakfast all around us, but we were in our own little world, taking our time over an ancient food that was more varied in color, texture, aroma and flavor than I had ever imagined.

I was still sucking some creamy white lavender honey off a little coffee spoon, when Xavier suddenly reached into the cardboard box and pulled out an ordinary short-stemmed wine glass and set it down in front of me. Again from the box, and without a word, he pulled out a plastic lab bottle, with technical writing scrawled on the side. It brought to mind harsh chemicals and unpleasant smells, and was jarringly out of place in this sweet smelling setting. And from the bottle he poured a colorless liquid into the wine glass.

"This is the mead?" I asked.

"Yes," he said.

I picked up the glass, and put it down again. I was suddenly overwhelmed by a ridiculous shyness in finally coming face to face with it and an unexpected dread of having to taste it in front of him.

"You should try it before the aromas disappear," he said. He was as nervous as I was.

The wine glass sat on the table before me. I twisted the stem back and forth with my thumb and index finger. Be careful what you wish for, was all I could think.

"Is it the right temperature, or do I need to warm it up?" I asked, thinking about the banana beer I had tasted so long ago, but actually just stalling for time.

"It's fine," he said.

I hesitated, and then took a little sniff and finally took my first taste. The moment I had been imagining for over a year had finally arrived and not only was I flustered, I was baffled. It wasn't that I didn't like the taste—it was that I couldn't really *taste* much of anything. The liquid in the glass was as clear and still as tap water; it did not shimmer with other-worldliness, as it had in my imagination. The aroma eluded me and the taste did not overwhelm me—but what, after all, could have lived up to my inflated expectations? And then I saw the disappointment on Xavier's face.

"I think my taste buds are coated in honey," I said. I poured myself some of the mineral water, and drank. I tried the mead again, but nothing. I sipped again and again, and it was always the same. I have no memory of what I said to him, but I didn't have to say anything—Xavier didn't need a Restaurant of the Future "Face Reader" to know what I was thinking. I felt terrible—like a stranger who had invited herself to the table with no sense of how to appreciate what had been so carefully and lovingly prepared and set before her.

Xavier recovered, or perhaps hid his disappointment, and started to tell me about the mead.

It was three weeks old, but would eventually be aged anywhere from six months to a year. It was semi-dry, didn't have any sulfites, had some tannin. The alcohol content was

11 percent; the acidity and sweetness were to his liking. However, he wanted the alcohol content of the final product to be 12.5 percent.

"This is not a finished product," he said.

That explained a lot, I thought, but not everything. He also talked about future variations he was considering, such as mead made with fruit, as well as sparkling mead.

"Perhaps I will make my product in Champagne," he said, referring to the famous region in Northern France not far from the Belgian border. He took the glass out of my hand and took a sniff, and a sip.

"I'm very satisfied with the aroma," he said.

"What did Sonia think?" I asked.

"For her it was a new experience too," he said. "She loves it, but it needs structure. She loved the taste."

I finished what was in the glass, still trying to figure out why I hadn't gotten more out of it. And as I continued listening to him I absentmindedly held the empty glass to my nose.

"It smells like honey," I said.

Xavier brightened up, and handed me one of the pots of honey to sniff—one which he and Sonia had used for his mead. (Three different honeys had been used in the end: one for the alcohol, and two for flavor.) Back and forth I went between the empty glass and the pot of honey, sniffing and comparing, and delighted that the aromas were exactly the same.

I wondered why I could smell the aroma of the mead

better after I had drunk it, and the glass was empty. Then suddenly something occurred to me. The door leading to the baking rooms had never closed as we sat there because the waitress was still walking back and forth past our table with trays of tarts. It was difficult to smell and taste the mead with all of the competing aromas from the baking rooms and the passing tarts, but easier when I covered my nose with the empty glass.

Being unable to smell the mead had greatly diminished my ability to taste it. How tricky our perception of the world is! On the surface everything we're surrounded by is so concrete and seemingly ordered and straightforward. In reality, everything is so ephemeral, so quick to be influenced by other factors, or misinterpreted.

"No wonder I can't smell or taste much," I said, motioning to the open door.

I asked for another chance, another taste, and Xavier poured me another half a glassful. I stood up and faced away from the open baking room door, and took a drink. Next I sat a few seats down from where he was sitting, with my back to the door, and again drank. But before I could contort myself any further, Xavier stood up.

"Let's go outside," he said, and so we did.

We stood in front of the bakery, on the outside, looking in. The tempting pastry cases held rows of jewel-colored fruit tarts framed by a huge picture window at the front of the shop. Above the window hung hundreds of tiny fairy lights, strung together like tangled strands of bright pearls and draped from one end of the window to the other. Two golden Valentine's Day hearts at the center of the lights completed the picture. This was the prettiest store front in the square. It was raining lightly and Xavier made sure I stood close to the window, sheltered from the drizzle. The air was fresh and clean and cool, and I breathed deeply trying to cleanse my nostrils of the saturated scents of the bakery. I took another drink from the glass, and there it was—flavor. I tasted again. Coming outside had made a startling difference—as good a trick as the milk/water transformation of the "magic flavor," at IFF in the Netherlands: Remove the smell of the bakery, and the aroma and taste of the mead were free to emerge. I smiled, I was complimentary, I drank again, I offered Xavier a taste, but I'm not sure that he believed either my kind words or my enthusiasm.

I had come outside without my coat, and I was shivering, but it didn't matter, and again I sipped. The flavor was fragile, but infused with all of the delight and promise of a flower in bud. Perhaps the mead version of Beaujolais Nouveau. Water had been turned into honey wine, and would only get better and more complex over time. I was tasting Xavier's past, as well as a hint of some future pleasure. I took one last drink, and it was gone.

I realized the difficulties I had had during this surprise tasting session lay not entirely with the sweet smelling venue or the immature mead, but with me. I had been waiting more than a year to be presented with a single, immutable substance, but I now realized that Xavier—beekeeper, honey addict, mead maker, chef, entrepreneur and science enthusiast—was playing with flavors and aromas that would forever be in transition: Nectar becoming honey, honey becoming mead, and mead + roast leg of lamb or chocolate or other ingredients, through the alchemy of cooking, becoming something different, and potentially dazzling, altogether.

During my travels, I had discovered a microcosm of food and sensory science in my corner of Europe: a surprising and serious number of people, behind the scenes, and from every angle, involved in researching, creating, manipulating and maintaining flavor. And finally, with the taste of Xavier's mead, everything I had learned had come together. Flavor wasn't one thing, it was a multitude of things. Not like snow, singular, but like snowflakes, plural: Infinite and individual, childishly pleasant on the tongue, quick to fade away, and exquisite in recollection.

I had enjoyed myself hugely, starting at the beginning with Hughes and his wicked but charming way of snooping around people's kitchens in order to get to know them. Now that I had come to the end of my search, in which flavor had grown, out of proportion, to take on such a singular significance, I thought: imagine an exaggerated Hughes-like world where people got to know each other, not by going into their kitchens and eyeing their food stores, but by going into their kitchens and saying, "What are you cooking? Tell me about your life through your favorite foods and recipes; tell me your cooking secrets and your flavor memories." Not speed-dating, but slow food dating—cook for me, cook with me, let me cook for you—possibly more intimate, and certainly more substantial and lasting than a group aphrodisiac dinner. I suppose we have a tamer tradition of this already, if we are lucky enough to have large tables for lingering meals with family and friends, with flavor as the glue that holds the experience together and fixes it in our memories as a keepsake.

Xavier and I walked back into the bakery, and one of the women behind the counter stared openly at me as we headed back to our table. Going outside to smoke a cigarette, that wouldn't have raised eyebrows, but this was a country where food was serious business and I suppose I looked quite reckless and suspicious standing outside drinking a glass of "wine," and exposing it to the risk of being diluted with rain water.

We sat down at the table and Xavier began to reminisce about his love affair with honey, going back all the way to when he was nine and was friends with two boys at the end of his street who lived with their grandfather, a beekeeper. Xavier remembered vividly the first time he saw the honey extracted from the hives. And before he had even tasted it, the sight of it, sealed and packaged so neatly in individual cells, combined with the movement and sound of the hundreds of tiny female workers who had produced it, piqued his interest.

"I think at first I loved the bees more than the honey,"

he said. Through his friends' grandfather and his own early beekeeping experiences, he had virtually seen nectar become honey, but as he grew older, he would want to learn the secret of turning honey back into nectar.

"If I were intelligent enough, I would make mead directly from nectar," he said with a smile. He had resisted what Tantalus had not: stealing secrets from the gods.

We were interrupted by a call Xavier took on his phone from a beekeeper who was hoping to sell him honey for his mead. "I used to be the ones calling them, and now they're calling me," said Xavier, pleased with his growing reputation. He and Claire had just started a new company, Nectar & Co., which would sell a range of honey with unusual, subtle colors, textures, aromas and flavors, made from the blossoms of trees such as chestnut, eucalyptus, bergamot and lemon. And they were designing a website that would include videotaped segments on how to cook with mead: the first one up would be his lamb marinated in mead.

He had already given me the lamb recipe, along with another favorite mead recipe of his, Crumble of Christmas Boudin Sausage with Mead Sauce, based on boudin sausage from Wallonia—the southern, Francophone part of the country where Xavier grew up. It was a combination of tastes, textures and temperatures you could imagine being served at a cocktail party. Take a large balloon-type wine glass, and layer the ingredients: warm, crumbled white and black boudin sausage, warm apple compote, mead sauce, and crumbled *speculoos,* those Belgian spice cookies. The white boudins are made with milk; the black, with pig's blood. Xavier noted that boudins in Wallonia can vary in flavor—sweet or savory—depending on the imagination of the butcher who makes the sausage. Many years ago, Xavier wrote at the end of the recipe, the sausages were made when people in the region butchered their pigs, often around Christmastime, because that's when the pigs had the most fat, and because the meat would keep well in the cold weather. But these communal food production traditions have all but died out now: "The whole family, and often the whole village, would participate in this operation which ended with a group meal, much appreciated by the poor among them," Xavier wrote. A trophobiosis, of sorts: everyone pitched in to produce this traditional food, including stuffing the sausage casings, made of pig's intestines, in exchange for protection from hunger, at least for one day.

Along with Xavier's recipes, I was also taking away with me memories of the sensory experiences he had given me: the sound of baguettes singing; the aromas and textures of various types of honey; the sight of Au Vatel, in all seasons, and at different times of the day; the taste of his cooking, the taste of honey made from his own bees, and the first of many, I hoped, tastes of his new mead. I had given him nothing in return but a promise to make him some sopaipillas sometime and bring them to the bakery so we could taste them with some of the honey in his new range. I wondered how the taste of a warm New Mexican sopaipilla would change with, say, a spoonful of his tart, creamy lemon tree blossom honey from Italy.

I had learned something about flavor and science from everyone I had met. But the secret ingredient, the zest of the experience had been returning home and listening to each installment of Xavier's mead story sitting in this cozy, family bakery surrounded by the smell of bread, pastries and coffee. That's what helped blend the individual raw ingredients of my exploratory journey into a single, satisfying dish.

But not everything had come out even. I wondered what it would be like to revisit Magni's biscuit study in which unaccustomed flavor had kept the Ugandan children from the nutrition they needed. I wondered if anyone had ever picked up where she had been forced to leave off, and if the children, if they had survived, had children of their own now.

As Xavier packed up the cardboard box with the jars of honey, the wine glass, and the plastic lab bottle, I told him I was certain that after his mead was properly christened, attractively bottled, and cleverly marketed, it would be a sterling success. I hoped he believed me. After we said goodbye, I bought a jar of the Mount Etna honey for myself, as well as a tiny, pretty box of Valentine chocolates for my husband, and then I walked out of the sweet smelling bakery onto the old city square where the drizzle was entering that fuzzy stage right before raindrops turn into sleet.

Xavier still had more work and uncertainty ahead of him just to get the first bottles of mead out of the lab and into the stemmed glasses and saucepans of food lovers everywhere, and I admired his persistence. For most of his young life he had been tantalized by something so transitory, and had desired it so unceasingly that he could almost, but not quite, taste it. He was intelligent, hard-working, and driven, and I had no doubt that soon there would be hundreds of bottles of his mead on the wall, as the raucous old beer song goes, born of memory and science –still ones, sparkling ones, perhaps even fruit ones that would surely be infused with a certain ethereal liveliness that would set them apart from indifferently or mass-produced beverages. Xavier's future was a bright, sweet one. I could almost see him, eternally boyish, even into old age, forever chasing and bottling the honey-tinted hours of days past.

V. RECIPES

SOPAIPILLAS

Before leaving home in the late 1970s, I wrote down my mother's haiku-like recipe for sopaipillas on an index card: a list of ingredients with one instruction, "deep-fry." I've included some additional preparation methods here.

MAKES APPROXIMATELY 2 DOZEN

Ingredients
4 cups flour
4 teaspoons baking powder
1 teaspoon salt
4 teaspoons lard [or solid vegetable shortening]
warm water (about 1 cup)
vegetable oil for deep-frying

Note: You may have to experiment with the thinness of the dough, as well as the temperature of the hot oil, as you observe how well they puff up when you deep-fry them. They should puff up like little triangular or rectangular balloons, but those that don't inflate all the way are still delicious.

With a large spoon, mix the flour, baking powder, and salt in a mixing bowl. Add the lard or vegetable shortening and blend gently using the tips of your fingers.

Add warm water a little at a time until dough comes together, and then remove from bowl and knead on a lightly floured surface until smooth. Form the dough into a ball, cover, and let rest for an hour at room temperature.

Heat the vegetable oil in a heavy pan or deep fryer. Roll out half the dough, about as thin as a piece of sliced cheese on a lightly floured surface. Keep the other half of the dough covered to keep it from drying out. Cut rolled-out dough into triangles or rectangles, about 4 inches across.

Slip the dough pieces into the hot oil one at a time—you may get 2 or 3 in at once, depending on the size of your pan, but don't crowd them. They should start to puff up almost immediately. Turn once, letting them become golden brown on both sides. When browned on both sides, remove with a slotted spoon and drain on paper towel. They can be kept warm in the oven.

Serve in a bread basket, and let each person add their own honey.

Why do sopaipillas puff up?

I asked this of Lisa McKee, professor of human nutrition and food science at New Mexico State University, who explained, "It's a combination of heat and gluten and starch. Sopaipillas are typically fried, which is a high-heat method of cooking. This heat causes water in the dough to turn to steam fairly rapidly. The steam is trapped for a period of time in the dough and since steam has a large amount of energy, it forces the layers apart—thus the puffing. Cooking continues, though, past the point where the majority of the steam has been generated and lost. That continued application of heat then causes the gluten structure to set and the starch in the dough to gelatinize. When the sopaipilla is removed from the heat, the gluten hardens and the starch forms a gel structure, both of which keep the sopaipilla puffed even after cooking."

LAMB CHOPS MARINATED IN MEAD

When I first tasted Xavier's lamb marinated in mead at the Slow Food tasting, he made it with roast leg of lamb, but this recipe of his is more elegant, and better suited to a sit-down dinner.

SERVES 4

Ingredients for the chops and the accompaniment
12 lamb chops
butter, slightly salted
4 tablespoons olive oil
4 small shallots
8 cloves of garlic, cleaned, but not peeled
2.2 lbs potatoes
2 sprigs fresh rosemary, minced
salt and pepper

Ingredients for the marinade
2 cups mead
1 onion, chopped
2 pinches white pepper
2 celery stalks, chopped
1 sprig fresh thyme
¾ cup honey vinegar
4 cloves of garlic
2 cups water

3 whole cloves
1 pinch ground coriander
5 bay leaves
1 sprig fresh rosemary
3 tablespoons sherry vinegar
1 tablespoon honey

Preparation the evening before:
Mix all of the ingredients for the marinade in a heatproof bowl. In order not to alter the aromas of the honey, heat the marinade no higher than around 140°F in a double boiler. When the honey is dissolved, remove from heat and allow to cool.

Place the chops in a large casserole dish, and pour the marinade over the meat. Refrigerate, turn the meat periodically, and leave in the refrigerator overnight.

The next day:
Preheat oven to 410°F. Peel the potatoes, cut them into quarters, rinse and dry. Put them in a mixing bowl, and pour over 4 tablespoons olive oil; sprinkle with salt and freshly ground pepper. Add the shallots (cut into 4), garlic, and rosemary. Mix well, and put the potatoes in a roasting pan. Roast for approximately 40 minutes in the preheated oven. Stir from time to time and lower the oven temperature if necessary.

Remove the chops from the marinade. Strain the marinade, through a wire sieve, into a saucepan, and let it reduce over a low heat. When it's time to serve, mix in a knob of butter.

Cook the chops in a little olive oil for 4-8 minutes, depending on their thickness, until they are cooked the way you like them—from rare to well done. Sprinkle with salt and freshly ground pepper. Keep warm before serving.

To serve:
Place approximately 2 spoonfuls of the potatoes at the center of each of the four dinner plates to form a base for the chops. Arrange three chops per plate on the potatoes, bone-end to the center, so that they form a point in the middle.

Drizzle the chops with the marinade that has been transformed by slow cooking and the addition of butter into mead-flavored *oignons confits.*

Thanks to Xavier Rennotte.

FLOWER POWER ELDERFLOWER COCKTAIL

SERVES 1

Serge Guillou, a Frenchman and president of Belgium's Bartenders Guild, made this cocktail especially for this book at the Hotel Bloom in Brussels. He develops the cocktails served at the hotel's bar and restaurant, Smoods, which has seven dining areas, or "mood islands" as they call them, which reminded me of the mood rooms at the Restaurant of the Future. Serge and I decided elderflower should be the featured flavor of this new cocktail: a marriage of elderberry, the flavor of the first mead Xavier ever tasted, and flower, the signature motif of the hotel.

Pour directly into a champagne glass, in the order given:
1¼ tablespoons St-Germain elderflower liqueur
2 teaspoons Marie Brizard pear liqueur
1 tablespoon grapefruit-flavored flavor pearls
(*perles de saveur*)*

**available online or at gourmet food stores*

Fill the rest of the glass with Champagne—and watch the flavor pearls, or "molecules," swirl and dance in the liquid, and eventually settle on the bottom of the glass. Add

1 dash Fee Brothers Lemon Bitters and give the drink a single, gentle stir.

With thanks to Serge Guillou and Marion Flipse at Hotel Bloom.

What are bitters?

Joe Fee, one of the owners of Fee Brothers, a family business since 1864, told me, "bitters are a combination of extracts from roots, barks, spices, and fruits. It's our belief that they got their name because in their concentrated form they do taste bitter. But like vanilla extract, we all know that used properly they are anything but bitter."

HOP SHOOTS WITH POACHED EGG AND SMOKED SALMON

SERVES 1

Hop shoots are a seasonal delicacy in Belgium. I once paid around $110 per pound for them at a gourmet grocery store in Brussels where a big sign on the door announced, "Hop Shoots Have Arrived." This recipe was given to me by the Flemish chef Stefaan Couttenye, owner of the Hommelhof (Hop Garden) restaurant. To me, hop shoots look and taste similar to bean sprouts, but chef Stefaan thinks they have a hint of walnut flavor. Hop shoots go well with quite a few ingredients, but are at their best with just a poached egg.

Ingredients

1 egg
2½ ounces hop shoots
1 slice of smoked salmon
1 slice of fried bread
chervil
¼ cup fresh cream

Ingredients for the mousseline sauce

2 egg yolks
⅓ cup melted butter
squeeze of lemon juice
salt and pepper
nutmeg

Clean each hop shoot by breaking off the hard ends. The remaining shoot should be 2–3 inches long. Thoroughly rinse the shoots and blanche them for 2 minutes in lightly salted water to which a little lemon juice is added. They should remain crunchy.

Whisk together the ingredients for the mousseline sauce.

Heat the hop shoots in a little cream and season with salt, pepper, and nutmeg to taste.

To serve, place the hop shoots on a pre-heated dish and put a slice of smoked salmon on top. Poach the egg and put it carefully on the salmon. Pour over the mousseline sauce. Garnish with some chervil and serve with a slice of fried bread.

With kind permission of Stefaan Couttenye.

PEPPERMINT CHIP ICE CREAM

Mint has an extraordinary trigeminal cooling effect, and goes well with many foods, sweet and savory. My favorite combination is mint with chocolate—especially mint chocolate chip ice cream. This recipe is from Van Leeuwen Artisan Ice Cream, a New York-based company that sells from ice cream trucks and in its own shops.

Ingredients

1 teaspoon organic peppermint extract
¼ cup chopped 65–75% chocolate
¾ cup whole milk
1 cup heavy cream
⅓ cup sugar
4 egg yolks

Combine milk, cream, and sugar in a double boiler and heat the mixture to 125°F.

Whisk the egg yolks in a separate bowl. Slowly add about a quarter of the hot milk mixture to the egg yolks, whisking throughout to allow it to temper and to avoid scrambling the yolks.

Once mixed, pour the egg mixture into the double boiler with the remaining milk mixture. On a medium heat, bring the mixture up to 170°F and hold it there for about 4 minutes or until it holds its form. (To test for this stage, coat a wooden spoon with the mixture and run your finger over the spoon; if the mark from your finger holds its form, the mix has finished cooking.)

Transfer the mixture to a stainless steel bowl and place it in an ice bath in order to arrest the cooking process. Once cooled, add the mint extract. Put the mixture into the refrigerator overnight, or for at least 3 hours, to thicken.

Churn ice cream in an ice cream maker until the mixture gets to a soft serve consistency, then gently fold in chocolate chips. If you don't have an ice cream maker, you can use an electric mixer with a whisk attachment on the lowest speed to "churn" the ice cream. Transfer to a freezer container and freeze until hard.

Many thanks to Laura O'Neill, co-owner and founder of Van Leeuwen Artisan Ice Cream.

APPLE CHUTNEY

Ingredients

3–4 cooking apples
4 medium onions
¾ cup raisins
1¾ cups vinegar
2 cups sugar
1 teaspoon salt
1 teaspoon curry powder
½ teaspoon mustard
½ teaspoon pepper
½ teaspoon ground ginger

Peel and core the apples and peel the onions and chop into small pieces. Mix all the ingredients except the sugar and boil gently until soft. Add the sugar and boil for 30 minutes.

This recipe is from *The Composition of Foods,* by Robert Alexander McCance and Elsie Widdowson, published by the Royal Society of Chemistry; 6th revised edition, 2002. Permission to reprint granted under the Open Government License of the United Kingdom.

The Composition of Foods by Robert Alexander McCance and Elsie Widdowson, first published in England in 1940, is a classic scientific reference book that also contains a number of traditional British recipes such as Steak and Kidney Pie, Yorkshire Pudding, and this recipe for apple chutney. According to its authors, knowledge of the chemical composition of food was essential for the dietary treatment of disease and the study of human nutrition. Co-author Elsie Widdowson (1906–2000), a chemist who began her scientific career with a study of the carbohydrate content of developing apples, was one of the first women to graduate from Imperial College, London, and received many honors in her lifetime, including Commander of the British Empire and Fellow of the Royal Society.

SABAYON WITH MUSA LOVA BANANA LIQUEUR

The Laboratory of Tropical Crop Improvement at Leuven University, in Flanders, holds the largest *Musa* (banana plant) collection in the world. This recipe is based on a new banana liqueur called Musa Lova, developed by local restaurant owner Fabian Deckers, with the help of Professor Rony Swennen, the director of the university's more than 1,000 Musa varieties. Musa Lova is made with rum, Cavendish bananas, honey from Leuven, and lime and lemon zest. The banana is one of the most important staple foods in the world—and a portion of Musa Lova's profits are donated to a banana development program in the Congo headed by Professor Swennen.

SERVES 6

Ingredients
4 egg yolks
4 egg whites
½ cup fine white sugar
2 cups Musa Lova with Honey
½ cup cream
1 gelatin leaf or three teaspoons powdered gelatin
juice of 2 limes
raspberries
6 small pieces of meringue
6 sprigs fresh mint

Whip the cream until it forms soft, not stiff, peaks. In a separate bowl, whisk the egg whites.

Put the egg yolks, sugar, Musa Lova liqueur and lime juice in a saucepan. Whisk the mixture over a medium heat until it becomes creamy and add the gelatin.

Remove from the stove and add the cream. Then carefully fold in the whisked egg whites to give some extra airiness.

To serve:
Cover the bottom of six wine glasses with raspberries. Add the sabayon on top of the raspberries. Chill for at least 4 hours.

Before serving, top with a few broken pieces of meringue, a few more raspberries, and a sprig of mint.

Many thanks to chef Wannes Ickx and Fabian Deckers.

DUTCH HAM AND ENDIVES GRATINÉES

Endive, *chicon, witloof*—it doesn't matter what you call this bitter, leafy vegetable, it can be eaten raw, as in a mixed salad, or cooked, as in this recipe. This dish calls for either endive or asparagus—both of which grow in abundance in Belgium.

SERVES 4

4 endives or 12 stalks parboiled asparagus
1 teaspoon salt
2 tablespoons butter
2 tablespoons flour
1 cup milk
1 cup grated Gruyere
2 tablespoons grated Parmesan
salt and pepper
nutmeg
4 slices boiled ham

Clean the endives and place them in a large skillet. Cover them with water, add the teaspoon of salt, and bring to a boil. Lower the heat, cover, and simmer for 5 minutes. Drain the endives and let them cool. Gently squeeze out as much liquid as possible.

Melt the butter in a saucepan. Add the flour and stir over low heat 3-5 minutes. Pour in the milk and whisk until the sauce is thick and smooth. Remove the saucepan from the heat. Add half the Gruyere and all the Parmesan, stirring until they have melted completely. Season with salt, pepper, and nutmeg.

Lightly butter a shallow baking dish and coat the bottom with a little of the cheese sauce. Wrap each endive (or three stalks parboiled asparagus, drained) in a piece of ham so that the vegetable shows at each end. Arrange the bundles in the baking dish, with the seam side of the ham down.

Spoon on the rest of the sauce and sprinkle with the remaining Gruyere. Bake at 350° F for 20-30 minutes, until the sauce is bubbling and the top is golden brown. If necessary, place under the boiler for a few minutes.

With kind permission of Barbara Wheaton, author of Savoring the Past: The French Kitchen and Table from 1300 to 1789, *and honorary curator of the Schlesinger Library's Culinary Collection at Radcliffe College.*

CRUMBLE OF CHRISTMAS BOUDIN SAUSAGE WITH MEAD SAUCE

This is another one of Xavier's favorite mead dishes. He says, "During Christmastime in Wallonia, butcher shops' windows are overflowing with boudin made with raisins, apples, walnuts, leeks, pumpkin, truffles, or Port . . . each butcher competing—with joy and imagination—to offer their clients a selection of sweet and savory boudin sausage."

SERVES 4

Ingredients for the boudin mixture
⅓ lb. white boudin with pecans
¼ lb. black boudin with raisins
a knob of butter

Ingredients for the apple compote
2 cooking apples
2 tablespoons sugar
¼ cup water

Ingredients for the mead sauce
2 cups veal stock
1¼ cups mead

2 oz. store bought speculoos cookies

Preparation ahead of time:
Prepare the compote the day before or in the morning, so that it can be well chilled before serving.

Peel and cut the apples into chunks. Cook the apples in the water on a high heat. After 5 minutes, mash the apples, drain off any excess water, and add the sugar. Chill.

Preparation before serving:
Remove the skin of the sausages, and place the meat in a mixing bowl. Mash the sausage meat with a fork.

Cook the sausage meat in the butter in a nonstick pan on a high heat. Remove when the meat is browned, and keep warm.

To create the mead sauce, combine the veal stock and the mead in a saucepan, simmer, and reduce. Salt and pepper to taste.

Prepare the speculoos cookies by breaking them into small pieces.

To Serve:
In 4 balloon-type wine glasses, layer the ingredients in the following order:

2 tablespoons warm sausage meat
1 tablespoon mead sauce
2 tablespoons cold compote
1 tablespoon crumbled speculoos cookies

Thanks to Xavier Rennotte.

MADELEINES

Ingredients

¼ cup + 1 teaspoon salted butter
1 medium egg
¼ cup + ½ teaspoon caster sugar
¼ cup minus 1 teaspoon plain flour
¼ cup minus 2 teaspoons ground almonds*
1 lemon, zested

a Madeleine baking pan

MAKES 1 DOZEN

**To avoid nuts, substitute the same amount flour.*

Preheat the oven to 375°F. Warm a heavy saucepan over a moderate heat and add the butter. Cook the butter slowly until it has melted, turned a golden color, and gives off a nutty scent; hence the name "*beurre noisette.*" Remove from the heat and allow to cool slightly.

In a metal bowl, whisk the egg with the caster sugar until the mixture has become light and airy. You should be able to briefly leave a figure eight with the whisk on the surface of the mixture.

Sift the flour and ground almonds into the bowl and gently fold into the egg mixture together with the lemon zest.

Gently stir the *beurre noisette* through the mixture. Leave to rest for about an hour if you have the time, to allow the gluten in the flour to relax, ensuring the cakes are light.

Spoon the batter into the madeleine molds filling them 3/4 full. Bake in the preheated oven for 8-10 minutes or until golden brown and springy to the touch.

Leave to cool in the mold until cool enough to handle and then turn out onto a wire rack. Once cooled, store in an airtight container.

With kind permission of Michael Vanheste, Belgian-born tutor at Bettys Cookery School in Harrogate, North Yorkshire, England.

What does* beurre noisette *sound like?
This recipe was featured on BBC Radio 4's *Woman's Hour* program, during which Michael Vanheste explained how to make the perfect *beurre noisette* (hazelnut butter). In the beginning the butter starts to melt and sizzle, and the fat separates from the milk solids. Listen to your butter, he says. As the bubbles diminsh, you'll see brown sediment at the bottom of your pan, your pan will go quiet, and you'll know it's done.

For more recipes related to the foods and flavors explored in this book, as well as more information on prominent women scientsts who have made significant contributions to the study and practice of the science of food, please visit

TASTEOFMOLECULES.COM

ACKNOWLEDGMENTS

Many thanks to Gloria Jacobs, Florence Howe, and everyone at the Feminist Press for their kindness and support. A huge thank you to all of the scientists who shared their stories with me, and to the chefs, food historians, and others who gave me their recipes to accompany some of the foods and flavors explored in the book—*et merci* to Janis Bergraat, owner/chef of Tastefully Seasoned Catering, for testing some of those recipes. And special thanks to Xavier Rennotte—beekeeper, chef, and science enthusiast whose search for a certain, extraordinary nectar was the aromatic glue that held all of the other stories together. I would also like to mention the media company, Science|Business, for helping me to discover, long before I even began this project, that science and technology—in education, policy, and industry—need to be as transformative, engaging, and accessible today as they were millions of years ago when the first humans used stone chopping tools as the world's earliest food processors. And, finally, to my husband and our two daughters: You are my past and my future; you are the best part of my memories and my dreams. But the fullest, most satisfying moment of any given day is lingering at the dinner table, listening to your stories, even after all the food is gone.

The Feminist Press is an independent, nonprofit literary publisher that promotes freedom of expression and social justice. Founded in 1970, we began as a crucial publishing component of second wave feminism, reprinting feminist classics by writers such as Zora Neale Hurston and Charlotte Perkins Gilman, and providing much-needed texts for the developing field of women's studies with books by Barbara Ehrenreich and Grace Paley. We publish feminist literature from around the world, by best-selling authors such as Shahrnush Parsipur, Ruth Kluger, and Ama Ata Aidoo; and North American writers of diverse race and class experience, such as Paule Marshall and Rahna Reiko Rizzuto. We have become the vanguard for books on contemporary feminist issues of equality and gender identity, with authors as various as Anita Hill, Justin Vivian Bond, and Ann Jones. We seek out innovative, often surprising books that tell a different story.

See our complete list of books at **feministpress.org**, and join the Friends of FP to receive all our books at a great discount.